Experiments Manual

to accompany

Electricity
Principles & Applications

Ninth Edition

Richard J. Fowler
Mitchel E. Schultz

Mc Graw Hill

EXPERIMENTS MANUAL TO ACCOMPANY ELECTRICITY PRINCIPLES AND APPLICATIONS, NINTH EDITION

Published by McGraw Hill LLC, 1325 Avenue of the Americas, New York, NY 10019.

This book is printed on acid-free paper.

1 2 3 4 5 6 7 8 9 LKV 27 26 25 24 23 22

ISBN 978-1-264-27074-3
MHID 1-264-27074-7

Cover image: *Blue Andy/Shutterstock*

mheducation.com/highered

Contents

Chapter 7 Magnetism and Electromagnetism 77

Chapter 8 Alternating Current and Voltage 91

Chapter 9 Power in AC Circuits 105

Chapter 10 Capacitance 117

Chapter 11 Inductance 137

Chapter 12 Transformers 159

Chapter 13 *R*, *C*, and *L* Circuits 177

Chapter 14 Electric Motors 199

Preface

This experiments manual contains a wide variety of tests and experiments designed to aid students in understanding electrical principles and applications. The structure and content of this experiments manual parallel the theory presented in the companion textbook, *Electricity: Principles and Applications, Ninth Edition*. Students using this manual will find that working with basic circuits will enhance their understanding of the theory.

This manual contains many troubleshooting experiments. Some of these experiments can be performed by using real components and instruments or by using Multisim. Multisim is an interactive circuit simulation software package that allows the user to view their circuit in schematic form while measuring various parameters of the circuit. The ability to quickly create a schematic diagram and then analyze the circuit through simulation makes Multisim a wonderful tool to help students understand the concepts covered in the study of electricity and electronics. Some experiments are designed specifically or exclusively for Multisim. Many of the other experiments can also be completed using either real components or Multisim.

In this edition, the experiments presented in the first seven chapters are primarily concerned with direct current, and those in Chaps. 8 through 13 deal with alternating current. Chapter 14 covers ac and dc motor characteristics, Chap. 15 covers instruments for measuring dc and ac quantities, and Chap. 16 deals with residential wiring. Together, these 16 chapters constitute a complete, integrated coverage of basic electricity. Although not all the experiments in each chapter need to be used in the classroom, it is recommended that the chapters be covered in sequential order.

The manual also includes a selected group of advanced problems. All of the experiments and advanced problems will enhance and foster skills used by electronic technicians. The advanced problems may be chosen to serve the needs of a particular class or group of students, as well as the physical facilities available (such as shops and labs). The parts required for the experiments are standard materials available from many different hardware and electronic parts distributors. Whenever possible, the same components are used in more than one experiment. All of the experiments and advanced problems have been thoroughly classroom-tested.

Students are encouraged to see beyond the exercise of recording data, drawing lines, and creating curves. The collected data serves to reinforce the classroom theory, and laboratory experience also demonstrates the proper and practical use of test equipment. Students are encouraged to look and think beyond each experiment for trends and consistency of theory presented.

Each chapter begins with a comprehensive test that covers the chapter's content. The test may be taken in one or two class sessions, depending upon class needs. It is suggested that even-numbered questions be given on one day, and odd-numbered questions on the next. This approach results in similar content coverage on both days. This same type of split is recommended when both a pretest and a posttest for the chapter are desired. If the textbook and this manual are studied simultaneously, the test may be given after each chapter is completed in both the text and the experiments manual.

The experiments manual concludes with two appendixes, the first covering breadboarding, and the second outlining the purpose, value and general procedure for using a lab notebook.

Every effort has been made to develop experiments that support modern theory and current practice. Comments from both instructors and students are again most welcome.

Richard J. Fowler
Mitchel E. Schultz

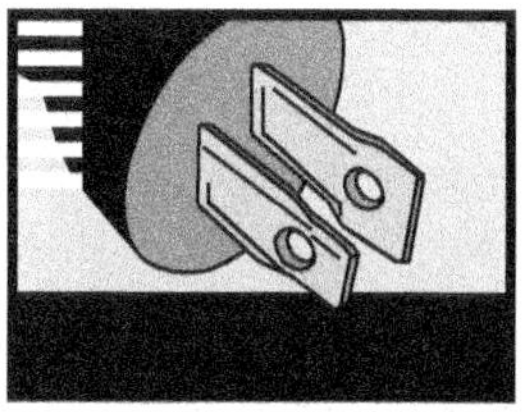

Safety

Electric and electronic circuits can be dangerous. Safe practices are necessary to prevent electrical shock, fires, explosions, mechanical damage, and injuries resulting from the improper use of tools.

Perhaps the greatest hazard is electrical shock. A current through the human body in excess of 10 milliamperes can paralyze the victim and make it impossible to let go of a "live" conductor or component. Ten milliamperes is a rather small amount of current flow: It is only *ten one-thousandths* of an ampere. An ordinary flashlight can provide more than 100 times that amount of current!

Flashlight cells and batteries are safe to handle because the resistance of human skin is normally high enough to keep the current flow very small. For example, touching an ordinary 1.5-V cell produces a current flow in the microampere range (a microampere is one-millionth of an ampere). This amount of current is too small to be noticed.

High voltage, on the other hand, can force enough current through the skin to produce a shock. If the current approaches 100 milliamperes or more, the shock can be fatal. Thus, the danger of shock increases with voltage. Those who work with high voltage must be properly trained and equipped.

When human skin is moist or cut, its resistance to the flow of electricity can drop drastically. When this happens, even moderate voltages may cause a serious shock. Experienced technicians know this, and they also know that so-called low-voltage equipment may have a high-voltage section or two. In other words, they do not practice two methods of working with circuits: one for high voltage and one for low voltage. They follow safe procedures at all times. They do not assume protective devices are working. They do not assume a circuit is off even though the switch is in the OFF position. They know the switch could be defective.

Even a low-voltage, high-current-capacity system like an automotive electrical system can be quite dangerous. Short-circuiting such a system with a ring or metal watchband can cause very severe burns—especially when the ring or band welds to the points being shorted.

As your knowledge and experience grow, you will learn many specific safe procedures for dealing with electricity and electronics. In the meantime:

1. Always follow procedures.
2. Use service manuals as often as possible. They often contain specific safety information. Read, and comply with, all appropriate material safety data sheets.
3. Investigate before you act.
4. When in doubt, *do not act.* Ask your instructor or supervisor.

General Safety Rules for Electricity and Electronics

Safe practices will protect you and your fellow workers. Study the following rules. Discuss them with others, and ask your instructor about any you do not understand.

1. Do not work when you are tired or taking medicine that makes you drowsy.
2. Do not work in poor light.
3. Do not work in damp areas or with wet shoes or clothing.
4. Use approved tools, equipment, and protective devices.
5. Avoid wearing rings, bracelets, and similar metal items when working around exposed electric circuits.
6. Never assume that a circuit is off. Double-check it with an instrument that you are sure is operational.
7. Some situations require a "buddy system" to guarantee that power will not be turned on while a technician is still working on a circuit.
8. Never tamper with or try to override safety devices such as an interlock (a type of switch that automatically removes power when a door is opened or a panel removed).
9. Keep tools and test equipment clean and in good working condition. Replace insulated probes and leads at the first sign of deterioration.

10. Some devices, such as capacitors, can store a *lethal* charge. They may store this charge for long periods of time. You must be certain these devices are discharged before working around them.
11. Do not remove grounds, and do not use adaptors that defeat the equipment ground.
12. Use only an approved fire extinguisher for electrical and electronic equipment. Water can conduct electricity and may severely damage equipment. Carbon dioxide (CO_2) or halogenated-type extinguishers are usually preferred. Foam-type extinguishers may also be desired in *some* cases. Commercial fire extinguishers are rated for the type of fires for which they are effective. Use only those rated for the proper working conditions.
13. Follow directions when using solvents and other chemicals. They may be toxic or flammable or may damage certain materials such as plastics. Always read and follow the appropriate material safety data sheets.
14. A few materials used in electronic equipment are toxic. Examples include tantalum capacitors and beryllium oxide transistor cases. These devices should not be crushed or abraded, and you should wash your hands thoroughly after handling them. Other materials (such as heat shrink tubing) may produce irritating fumes if overheated. Always read and follow the appropriate material safety data sheets.
15. Certain circuit components affect the safe performance of equipment and systems. Use only exact or approved replacement parts.
16. Use protective clothing and safety glasses when handling high-vacuum devices such as picture tubes and cathode-ray tubes.
17. Don't work on equipment before you know proper procedures and are aware of any potential safety hazards.
18. Many accidents have been caused by people rushing and cutting corners. Take the time required to protect yourself and others. Running, horseplay, and practical jokes are strictly forbidden in shops and laboratories.

Circuits and equipment must be treated with respect. Learn how they work and the proper way of working on them. Always practice safety: your health and life depend on it.

Engineer assembling laptop electronics.
Hero Images/Image Source

NAME ________________________________ DATE ______________

CHAPTER 1

Basic Concepts

TEST: BASIC CONCEPTS

For questions 1 to 9, determine whether each statement is true or false.

1. Many energy-conversion processes are 100 percent efficient. **1.** ______________

2. Neutrons and protons are contained in the nucleus of the hydrogen atom. **2.** ______________

3. The nucleus of any atom has a positive charge. **3.** ______________

4. Negatively charged particles repel each other. **4.** ______________

5. Atoms which have five or more valence electrons become negative ions when ionized. **5.** ______________

6. Protons have a larger diameter than electrons do. **6.** ______________

7. An element is composed of at least two types of atoms. **7.** ______________

8. An electric lamp converts electric energy to heat energy and light energy. **8.** ______________

9. The symbol (or abbreviation) for energy is J. **9.** ______________

For questions 10 to 17, choose the letter that best completes each statement.

10. Energy is defined as **10.** ______________
- **a.** A force in newtons
- **b.** A weight in kilograms
- **c.** The ability to do work
- **d.** A negative charge

11. The base unit of energy is the **11.** ______________
- **a.** Watt
- **b.** Joule
- **c.** Newton
- **d.** Nucleus

12. All matter is composed of **12.** ______________
- **a.** Atoms
- **b.** Compounds
- **c.** Ions
- **d.** Newtons

13. A generator that is turned by a mechanical device **13.** ______________
- **a.** Creates electric energy
- **b.** Converts heat energy to electric energy
- **c.** Converts electric energy to heat energy
- **d.** Converts mechanical energy to electric energy and heat energy

14. An object that possesses negative static charge has **14.** ______________
- **a.** An excess of electrons
- **b.** A deficiency of atoms
- **c.** An excess of neutrons
- **d.** A deficiency of protons

15. Which of the following electrons of a silver atom will be at the highest energy level (possess the most energy)? 15. ____________
   a. An electron in the first shell
   b. An electron in the second shell
   c. An electron in the valence shell
   d. An electron that has just escaped the valence shell

16. A positive ion is an atom that has 16. ____________
   a. Gained a valence electron
   b. Lost a valence electron
   c. Gained a proton
   d. Lost a proton

17. A lightning bolt is visible because 17. ____________
   a. Protons are jumping from atom to atom
   b. An air column is ionizing and deionizing
   c. Electric energy is being converted to heat energy
   d. Heat energy is being converted to electric energy

For questions 18 to 20, solve each problem.

18. How much work is done when an object is moved 2.1 m by a force of 1.8 N?

18. **Given:** ____________
____
Find: ____________
Known: ____________
Solution: ____________
Answer: ____________

19. What is the efficiency of a portable radio receiver that uses 900 J of energy from its battery to produce 190 J of sound energy?

19. **Given:** ____________
____
Find: ____________
Known: ____________
____
Solution: ____________
____
Answer: ____________

20. A device is 37 percent efficient and requires an input of 10,867 J of energy. What is its output energy?

20. **Given:** ____________
____
Find: ____________
Known: ____________
____
Solution: ____________
____
Answer: ____________

For questions 21 to 24, supply the missing word or phrase in each statement.

21. Inefficiency in a light bulb is caused by electric energy being converted to ____________ energy. 21. ____________

22. A precipitator uses ____________ electricity to remove dust from an air stream. 22. ____________

23. An object with a positive charge can induce a(n) ____________ charge on the surface of an uncharged object. 23. ____________

24. If an object with a negative charge touches an uncharged object, the uncharged object will obtain a(n) ____________ charge. 24. ____________

NAME ______________________ DATE ____________

1-1 LAB EXPERIMENT: STATIC CHARGES

PURPOSE

This experiment is designed to familiarize you with some of the characteristics of static charges. As an option, this experiment may be more suitable as a classroom or laboratory demonstration led by an instructor.

MATERIALS

Qty.

1 rod (glass or plastic) approximately 12 in. long and ⅛ to ½ in. in diameter
1 wire (magnet or hookup), 18 to 22 gage, about 14 in. long
1 ½-in. square of notebook paper or newspaper
1 thread, lightweight sewing, approximately 10 in. long

Qty.

1 piece of tape (masking, electrical, or transparent) approximately 1½ in. long
1 piece of cloth (silk, wool, or polyester) or leather about 10 × 10 in.
1 piece of Styrofoam cup approximately ¼ in. wide and 1 in. long

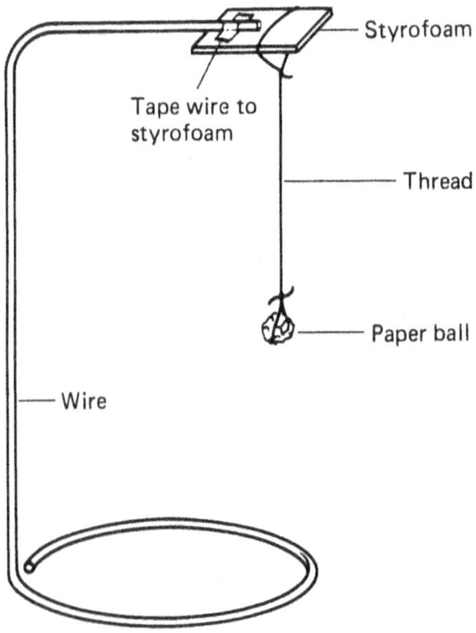

Fig. 1-1 Setup for checking static charge.

1. Bend the magnet or hookup wire into a stand as illustrated in Fig. 1-1. Set the stand on your table or desk. Crumple the ½-in. square of paper into a ball and tie the thread around it. Tie the other end of the thread around one end of the strip of Styrofoam. Tape the other end of the Styrofoam to the end of the wire stand (Fig. 1-1). Be certain that the tape does not touch the thread. Adjust (bend) the top of the stand until the ball clears the tabletop by at least 1 in.
2. Stroke the glass rod against the cloth or leather two or three times. Then slowly bring the end of the rod toward the free-hanging paper ball. Observe that the paper is attracted to the rod and makes contact with the rod; then the paper is quickly repelled (pushed away) from the rod. (Note: After being repelled, the paper ball may again be attracted to the rod if the charge on the rod is not very strong. If the charge on the rod is too weak, it may reattract and hold the paper for long periods of time.)

a. What causes the paper to be initially attracted to the rod?

____

____

____

b. What causes the paper to be repelled by the rod?

____

____

____

3. Tie another paper ball and thread to the Styrofoam (Fig. 1-2). Adjust the length of the new thread so that the two papers are hanging side by side. Separate the two threads just far enough so that the two paper balls are not touching.

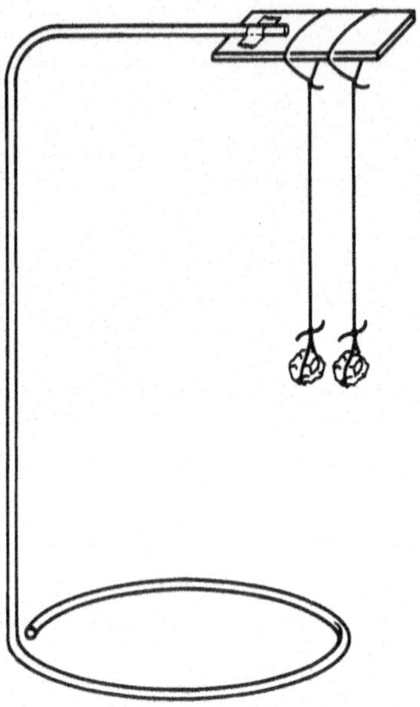

Fig. 1-2 Setup for checking the reaction between two charged objects.

4. Again charge the glass rod by stroking it two or three times against the cloth or leather. Now bring the rod toward the two hanging paper balls, and touch both of them at the same time. Then pull the rod down and away from the papers. (If the threads tangle up, straighten them and try again.) Note that the two papers repel each other.

a. Do the two balls have the same type of charge? ____________ Why?

____

____

b. How can a manufacturing industry use the principles you have just observed?

____

____

____

____

NAME ______________________________ DATE ______________

Electrical Quantities and Units

TEST: ELECTRICAL QUANTITIES AND UNITS

For questions 1 to 10, supply the missing word or phrase in each statement.

1. An electron possesses a(n) ______________ electric charge. 1. ______________
2. Ions can be current carriers in either ______________ or ______________. 2. ______________ ______________
3. The type of current that periodically reverses direction is ______________ current. 3. ______________
4. Energy in motion or being converted is ______________ energy. 4. ______________
5. A coulomb is the amount of charge possessed by ______________ electrons. 5. ______________
6. The type of materials used in insulators is classified as a(n) ______________. 6 ______________
7. Another name for resistivity is ______________. 7. ______________
8. The nonmetric unit for electric power is ______________. 8. ______________
9. ______________ describes a change in resistance caused by a change in temperature. 9. ______________
10. Atoms that have gained or lost electrons are called ______________. 10. ______________

For questions 11 to 37, choose the letter that best completes each statement.

11. The symbol for electric charge is 11. ______________
    - **a.** *C*
    - **b.** *W*
    - **c.** *Q*
    - **d.** *P*
12. "Coulomb" is the base unit for 12. ______________
    - **a.** Current
    - **b.** Power
    - **c.** Resistance
    - **d.** Charge
13. The abbreviation for coulomb is 13. ______________
    - **a.** W
    - **b.** C
    - **c.** Q
    - **d.** P
14. An electric current may be defined as the movement of 14. ______________
    - **a.** Protons through a copper conductor
    - **b.** Ions through a copper conductor
    - **c.** Charges in a specified direction
    - **d.** Charges in random directions

15. Current is represented by the letter

a. *E*
b. *I*
c. *C*
d. *A*

15. ____________

16. The watt is the base unit of

a. Current
b. Amperage
c. Voltage
d. Power

16. ____________

17. "A coulomb per second" is the definition of

a. An ampere
b. A watt
c. A joule
d. A volt

17. ____________

18. The abbreviation for the base unit of current is

a. W
b. V
c. C
d. A

18. ____________

19. Voltage can be defined as

a. Opposition to current
b. Potential energy difference
c. Movement of charges
d. Rate of doing work

19. ____________

20. "*P*" is the symbol for

a. Voltage
b. Power
c. Energy
d. Current

20. ____________

21. The voltage of a battery is available

a. At the negative terminal
b. At the positive terminal
c. Between the two terminals
d. At either the negative or the positive terminal

21. ____________

22. "Volt" is the base unit for

a. Wattage
b. Amperage
c. Voltage
d. Power

22. ____________

23. The base unit of voltage is defined as one

a. Ampere per second
b. Coulomb per second
c. Volt per second
d. Joule per coulomb

23. ____________

24. The abbreviation for volt is

a. W **c.** V
b. P **d.** A

24. ____________

25. Which of the following devices does not produce voltage?

a. Solar cell
b. Crystal
c. Conductor
d. Chemical cell

25. ____________

26. Resistance is defined as

a. Ability to conduct voltage
b. Ability to do work
c. Rate of doing work
d. Opposition to current

26. ____________

27. Electric energy is converted to heat energy by **27.** ________
a. Neutrons
b. Resistors
c. Insulators
d. Conductors

28. Which of the following materials is the best conductor? **28.** ________
a. Iron
b. Copper
c. Wood
d. Aluminum

29. The base unit of resistance is the **29.** ________
a. Joule
b. Ampere
c. Ohm
d. Coulomb

30. The symbol for resistance is **30.** ________
a. R
b. Q
c. P
d. Ω

31. The abbreviation for the base unit of resistance is **31.** ________
a. V
b. Ω
c. Q
d. R

32. "Rate of converting energy" is the definition for **32.** ________
a. Current
b. Kinetic energy
c. Work
d. Power

33. "Ampere" is the base unit for **33.** ________
a. Current
b. Work
c. Power
d. Voltage

34. "Joule per second" is a definition of the base unit for **34.** ________
a. Current
b. Work
c. Power
d. Voltage

35. The abbreviation for the base unit of power is **35.** ________
a. C
b. Ω
c. W
d. V

36. "V" is the symbol for **36.** ________
a. Voltage
b. Power
c. Current
d. Energy

37. Which of the following units is not a metric unit? **37.** ________
a. Horsepower
b. Coulomb
c. Second
d. Watt

For questions 38 to 43, solve each problem.

38. Determine the energy stored in a 12-V dry cell that has a charge of 2000 C.

38. Given: __________

____

Find: __________

Known: __________

Solution: __________

____

Answer: __________

39. How much current is flowing when a charge of 4 C, all moving in the same direction, passes a specified point in 2 s?

39. Given: __________

____

Find: __________

Known: __________

Solution: __________

____

Answer: __________

40. If $K = 0.004\ \Omega\cdot\text{cm}$ at 20°C, what is the resistance (at 20°C) of 250 cm of a solid conductor with a cross-sectional area of $0.02\ \text{cm}^2$?

40. Given: __________

____

Find: __________

Known: __________

Solution: __________

____

Answer: __________

41. A 500-W heater operates continuously for 3 h. How many watthours of energy does it convert?

41. Given: __________

____

Find: __________

Known: __________

Solution: __________

____

Answer: __________

42. What is the efficiency of an electric device requiring 51,600 joules every minute to produce 602 W of useful power output?

42. Given: __________

____

Find: __________

Known: __________

Solution: __________

____

Answer: __________

43. How much input power (in watts) is required by a motor that is 56 percent efficient and produces ¾ hp?

43. Given: __________

____

Find: __________

Known: __________

Solution: __________

____

Answer: __________

2-1 LAB EXPERIMENT: CONDUCTORS, INSULATORS, AND RESISTANCE

PURPOSE

In the following experiment, you will determine which materials are conductors and which are not. The last section of the experiment will help you visualize how physical dimensions determine resistance. As an option, this experiment may be more suitable as a classroom or laboratory demonstration led by an instructor.

MATERIALS

Qty.	
1	6-V battery (or other 6-V dc power supply)
1	no. 47 miniature lamp (6.3 V, 0.15 A)
1	lamp holder for no. 47 lamp (T-3¼ bayonet base)
3	test leads with alligator clips on both ends
1	teaspoon of sodium bicarbonate (baking soda)
2	15-cm (6-in.) lengths of bare (untinned) copper wire, 10 to 16 gage

Qty.	
1	Styrofoam cup
•	strips of various test materials such as paper, wood, copper, aluminum foil, glass, plastic, cloth, rubber, leather; the strips should be approximately 8 cm (3 in.) long; cross-sectional area is not important as long as connections can be made with alligator clips

PROCEDURE

(*Note:* To prolong the life of the battery, operate the light for as short a time as possible.)

1. Connect the battery, lamp socket, and wires, as shown in Fig. 2-1. Touch together the two alligator clips that will be used to make contact with the test materials. The lamp should light up to full brilliance. This indicates that the connecting wires are good conductors. Separate the alligator clips and the lamp should go off.

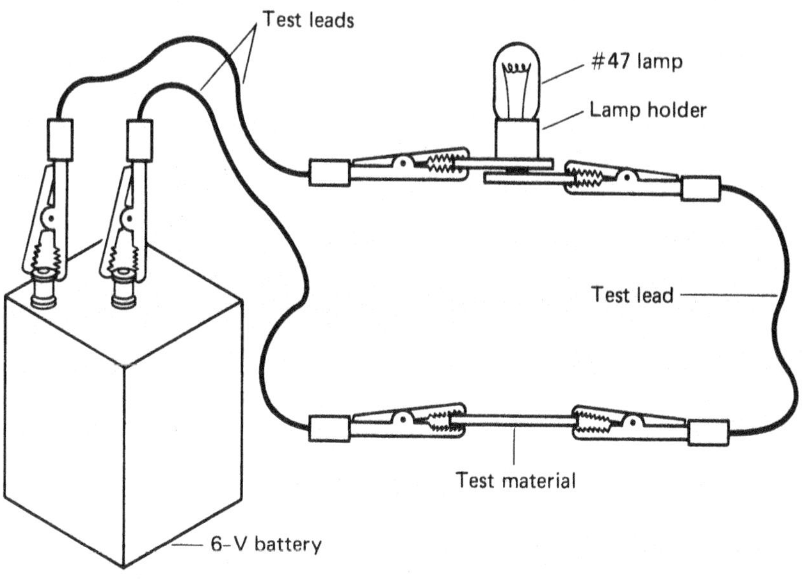

Fig. 2-1 Testing conductors and insulators.

2. Connect a piece of aluminum foil between the alligator clips and observe the brilliance of the lamp. Disconnect the alligator clips from the aluminum. Is aluminum a good conductor? ______________
3. Using the procedure outlined in step 2, test the other materials provided by your instructor. Record your results in Table 2-1. The first line of Table 2-1 is completed as an example.

TABLE 2-1 Insulators and Conductors

Material	Conductor (lamp lights)	Poor Conductor or Insulator (lamp does not light)
Aluminum	✓	
Paper		
Copper		
Wood		
Glass		
Plastic		
Cloth		
Rubber		
Leather		

4. Bend a piece of bare copper wire into the shape of copper wire 2 in Fig. 2-2. The end of the wire should be approximately 1 cm (⅜ in.) from the bottom of the cup. Connect the alligator clip from the lamp socket to this copper wire (Fig. 2-2). Connect the other piece of bare copper wire (copper wire 1 in Fig. 2-2) to the alligator clip that connects to the battery lead. Copper wire 1 and the alligator clip attached to it are held in place by hand when tests are being conducted. At other times, they can be placed on the table beside the cup.

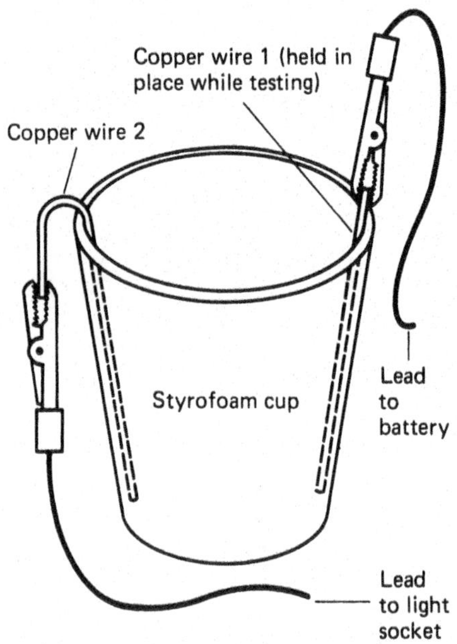

Fig. 2-2 Testing ionized solutions.

5. Fill the cup about two-thirds full of water.
   a. Insert copper wire 1 along the edge of the cup opposite copper wire 2. Does the lamp light? ______________
   b. Move copper wire 1 toward copper wire 2. Does the lamp light up before the two wires touch? ______________
   c. Is water a good conductor? ______________

6. Remove the alligator clip from copper wire 1. Add 1 teaspoon of baking soda to the water in the cup and stir with copper wire 1 (the baking soda will form ions when added to the water). Remove copper wire 1 from the solution and reconnect it to the alligator clip. Now slowly insert copper wire 1 into the solution along the edge opposite copper wire 2.
   - **a.** Does the light come on? ______
   - **b.** Is a baking soda solution a conductor? ______
   - **c.** While observing the lamp, slowly move copper wire 1 up and down in the solution. Does the light get brighter as the copper wire is moved into the solution? ______
   - **d.** With copper wire 1 fully inserted into the solution, slowly move copper wire 1 toward copper wire 2. Does the light get brighter as the wires are brought closer together? ______
   - **e.** Do you see any similarity between the depth that copper wire 1 was inserted into the solution and the cross-sectional area of a conductor? Why? (*Hint:* The current carriers—ions—move between the surfaces of the two copper wires.)

     ______

     ______
   - **f.** Do you see any similarity between the distance copper wire 1 was from copper wire 2 and the length of a conductor? Why?

     ______

     ______
   - **g.** Does increasing the cross-sectional area of a conductor increase or decrease its resistance? ______
   - **h.** Does increasing the length of a conductor increase or decrease its resistance? ______
7. Remove both wires from the solution. Dispose of the baking soda solution and the two copper wires in the manner specified by your instructor.

2-2 ADVANCED PROBLEMS

2-1. How much input power is required by a stereo amplifier that produces 100 joules per second and that is 65 percent efficient? ______________

2-2. What is the efficiency of a 1.0-hp motor that uses 6.1×10^6 J in 1 h? ______________

2-3. A 120-W stereo amplifier is 60 percent efficient. How many joules of energy does it use in 45 min? ______________

2-4. A 20-V battery has 2000 C of charge. It is connected to a 3.7-W electric device. How long will it take to use 50 percent of the battery's energy? ______________

2-5. A 14-gage conductor carries 15 A of current. How much charge, in a specified direction, passes a given point in the conductor in 4 s?

2-3 LAB EXERCISE: TEMPERATURE COEFFICIENTS OF RESISTORS

1. Using your computer, laptop or handheld device, search the Internet for electronic parts catalogs. (Examples would be Digi-Key or Mouser Electronics.) Under "Products", select or click on Resistors. If there are subheadings under resistors, look under precision, carbon-composition, metal-film, and wire-wound headings. Next, read the specifications for each type of resistor. You will not understand most of the specifications, but look for the temperature coefficient (tc) rating.
2. Record the highest and lowest temperature coefficients you can find. Also record the type of resistor (carbon, metal-film, etc.) that has the lowest and the highest temperature coefficient.

   ____

   ____

   ____

   a. Do most of the resistor listings include temperature coefficients? ______________

   b. Why do you think some resistor specifications do not include temperature coefficients?

   ____

   ____

NAME ______________________________ DATE ______________

CHAPTER 3

Basic Circuits, Laws, and Measurements

TEST: BASIC CIRCUITS, LAWS, AND MEASUREMENTS

For questions 1 to 14, determine whether each statement is true or false.

1. The metal frame of an automobile can serve as a conductor in its electric circuits. **1.** ____________

2. In an electric circuit, the load converts electric energy to another form of energy. **2.** ____________

3. The symbol for an insulated conductor is a solid line between two dashed lines. **3.** ____________

4. The physical arrangement of electric components is not indicated by a schematic diagram. **4.** ____________

5. Schematic diagrams never include the electrical specifications of the components. **5.** ____________

6. Current is directly proportional to resistance in an electric circuit. **6.** ____________

7. Doubling the resistance in a circuit halves the current if the voltage is held constant. **7.** ____________

8. Electron current leaves the positive terminal of the cell and travels through the load to the negative terminal of the cell. **8.** ____________

9. A digital panel meter is also known as a multimeter. **9.** ____________

10. A VOM is also known as a multimeter. **10.** ____________

11. The ohms-adjust setting on a VOM should be checked every time the resistance range is changed. **11.** ____________

Refer to Fig. 3-1 for questions 12 to 14.

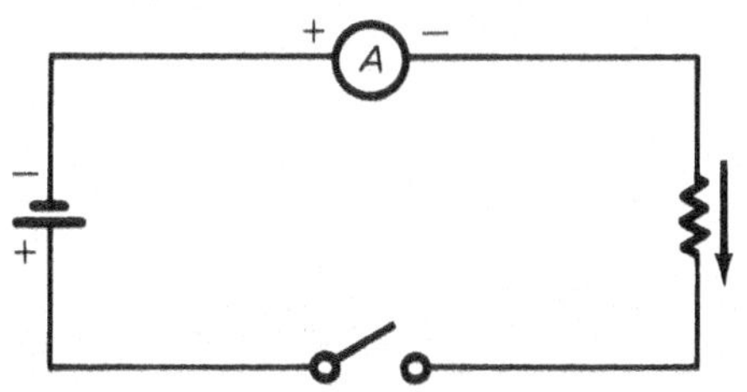

Fig. 3-1 Diagram for questions 12 to 14.

12. In Fig. 3-1, the arrow is pointing in the correct direction to indicate the direction of current (electron flow) when the switch is closed. **12.** ____________

13. The polarity shown on the meter is correct. **13.** ____________

14. The circuit contains a protection device. **14.** ____________

For questions 15 through 20, choose the letter that best completes each statement.

15. Which of the following is not an essential part of every circuit? **15.** ______
- **a.** Load **c.** Fuse
- **b.** Conductor **d.** Energy source

16. Which of the following is classified as a load for an electric circuit? **16.** ______
- **a.** Resistor **c.** Fuse
- **b.** Conductor **d.** Insulator

17. A circuit must be physically interrupted to measure **17.** ______
- **a.** Voltage **c.** Current
- **b.** Resistance

18. Power must be removed from a circuit when measuring **18.** ______
- **a.** Voltage **c.** Current
- **b.** Resistance

19. A multimeter cannot measure **19.** ______
- **a.** Power **c.** Voltage
- **b.** Current **d.** Resistance

20. Which of the following has no schematic symbol? **20.** ______
- **a.** Fuse **c.** Conductor
- **b.** Resistor **d.** Insulator

For questions 21 to 36, solve each problem. Be sure to include units (mA, V, etc.) in your answers.

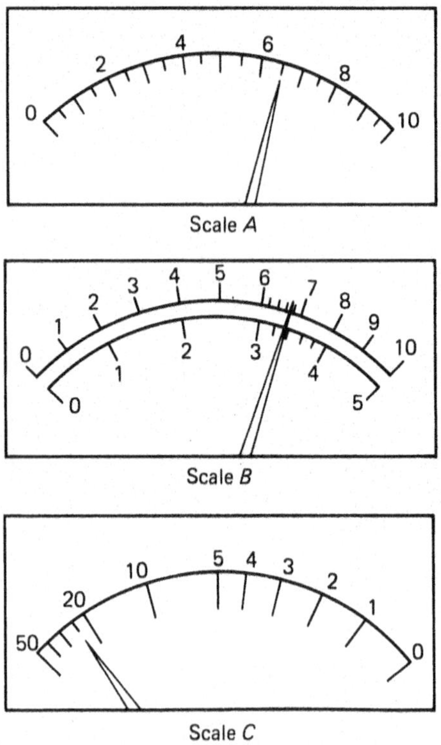

Fig. 3-2 Meter scales for questions 21 to 25.

21. Scale *A* in Fig. 3-2 represents part of a meter. How much voltage does the meter indicate if it is on the 100-V range? **21.** ______

22. Refer to scale *A* in Fig. 3-2. How much current does the meter indicate if it is on the 10-mA range? **22.** ______

23. Refer to scale *B* in Fig. 3-2. How much voltage is indicated if the meter is on the 1-V range? **23.** ______

24. Refer to scale *B* in Fig. 3-2. How much current is indicated if the meter is on the 5-mA range? 24. ______
25. Refer to scale *C* in Fig. 3-2. How much resistance is indicated if the meter is on the $\times 100\ \Omega$ range? 25. ______
26. How much current flows through a 40-Ω resistor that is connected to a 24-V battery? 26. ______
27. How much voltage is required to force 40 mA of current through a 100-Ω load? 27. ______
28. What is the resistance of a 12.6-V lamp that draws 1.4 A of current? 28. ______
29. What is the power of the lamp in question 28? 29. ______
30. How much energy is used by the lamp in question 28 if it operates for 6 h? 30. ______
31. Assuming that electric energy costs 9¢/kWh, how much does it cost to operate a 900-W toaster for 3 h? 31. ______
32. How much current is required by a 28-V, 40-W lamp? 32. ______
33. How much power is required by a circuit in which 5 A flows through a 15-Ω resistor? 33. ______
34. How much voltage is across a 5-kΩ resistor that has 2.5 mA of current flowing through it? 34. ______
35. If it costs 36¢ to operate a 60-W lamp for 50 h, what is the cost rate of energy? 35. ______
36. What is the resistance of a 40-W, 120-V lamp? 36. ______

NAME ______________________________ DATE ______________

3-1 LAB EXERCISE: WORKING WITH SCHEMATIC DIAGRAMS

PURPOSE

This lab will make you aware of the differences between the pictorial representation of a circuit and its schematic diagram.

1. Draw a schematic diagram of the pictorial presentations shown. Show the polarity of the battery and the meter on the diagram. Also indicate the direction of current by an arrow.
   a. Fig. 3-3
   b. Fig. 3-4
2. Refer to Fig. 3-5. Record on each diagram
   a. What is being measured
   b. The correct polarity of the meter if polarity is important for the indicated measurement
   c. The direction of current flow
3. Name each component in Fig. 3-5(*b*).
4. Would the meter reading in Fig. 3-5(*c*) change if the switch were opened?
   ______________ Why?
   ____
   ____
5. Would the meter reading in Fig. 3-5(*a*) change if the switch were opened?
   ______________ Why?
   ____
   ____
6. What might happen if the switch in Fig. 3-5(*d*) were accidentally closed?
   ____
   ____

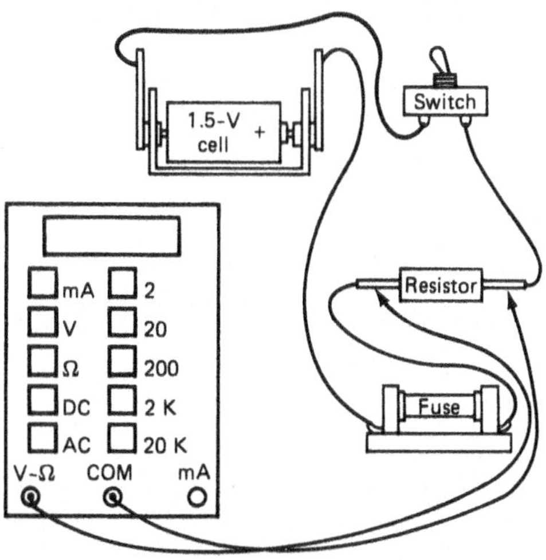

Fig. 3-3 Pictorial presentation for step 1a.

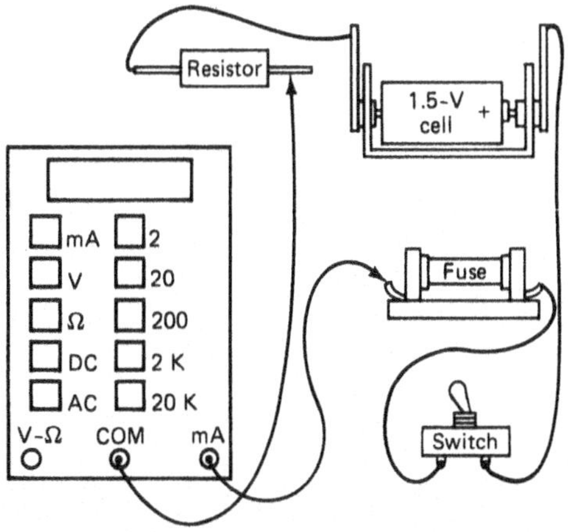

Fig. 3-4 Pictorial presentation for step 1b.

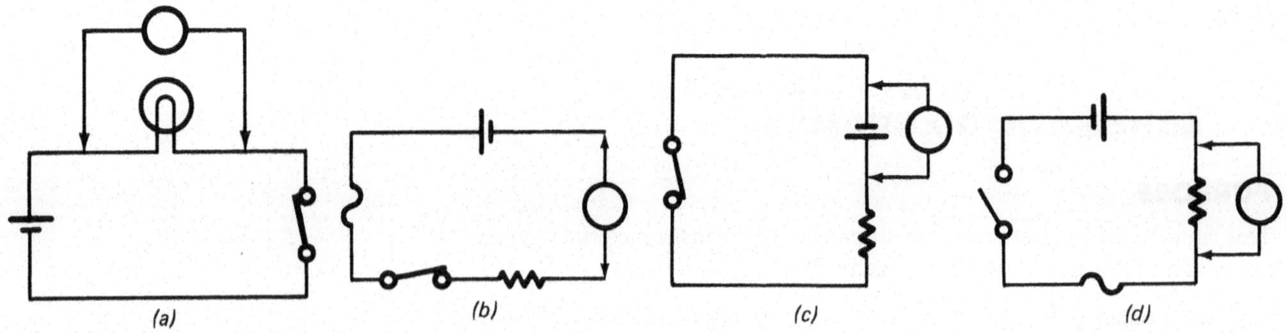

Fig. 3-5 Schematic diagrams for steps 2 to 6.

NAME ______________________________ DATE ______________

3-2 LAB EXPERIMENT: MEASURING VOLTAGE, CURRENT, AND RESISTANCE

PURPOSE

This experiment will provide practice in using the multimeter.

MATERIALS

Qty.		Qty.	
1	multimeter (DMM or VOM)	1	1.5-V dry cell
1	6-V dry battery (or other 6-V dc power supply)	1	1000-Ω, ½-W resistor, ±5%
		1	1-MΩ, ½-W resistor, ± 5%

PROCEDURE

1. Set the multimeter to the dc voltage function and the 10-V or higher range.

a. Observing polarity, measure and record the voltage of the cell and the battery. Cell ______________ Battery ______________

b. Next, select a range just larger than 1.5 V. Measure and record the voltage of the cell. ______________

c. Did measuring the cell voltage on a lower range yield a higher resolution? ______________ Why?

____

____

2. Set the multimeter to the resistance function. Select the 2-k range of the DMM or the ×10 range of the VOM. If using the VOM, short the leads together and ohms-adjust.

a. Measure and record the resistance of the resistor that is marked 1000 Ω. ______________

b. Change the meter to the 20-k or ×100 range. If using the VOM, short the leads together and ohms-adjust. Measure and record the resistance of the 1000-Ω resistor. ______________

c. Which range provides the most accurate measurement? ______________ Why?

____

____

d. Change to the 2-M or ×100-k range. Ohms-adjust if necessary. Measure the 1-MΩ resistor, being sure not to touch the metal tips of the test leads. Now, with your thumbs and forefingers, hold the metal tips of the test leads to the resistor leads. Does the resistance indicated on the ohmmeter change? ______________ Why?

____

____

3. Set the multimeter to the current function and the 10-mA range. Connect *all but one connection* of the circuit shown in Fig. 3-6. Have your instructor check your circuit before you make the last connection.

a. Make the last connection, and measure and record the current. ______________

b. Calculate and record the current for Fig. 3-6. ______________

c. Are the calculated and measured current values within 10 percent of each other? ______________ Should they be? ______________

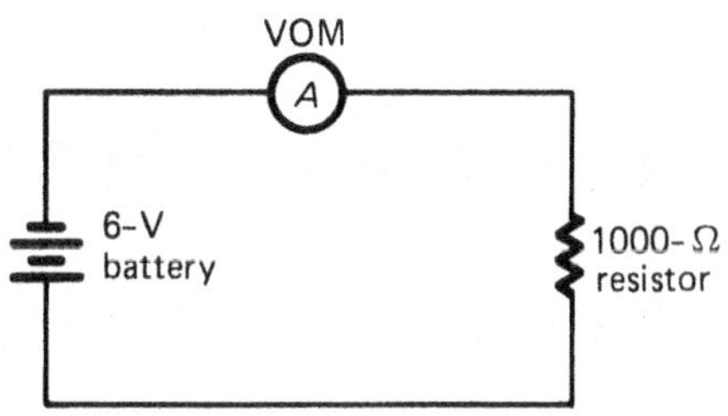

Fig. 3-6 Measuring current.

3-3 LAB EXPERIMENT: OHM'S LAW

PURPOSE

This experiment provides verification of Ohm's law. It also provides additional experience in constructing circuits and using meters.

MATERIALS

Qty.		Qty.	
1	multimeter (DMM or VOM)	1	1.5-V dry cell
1	6-V dry battery or power supply	1	1000-Ω, ½-W resistor
		1	10-kΩ, ½-W resistor

PROCEDURE

1. Refer to the circuit shown in Fig. 3-7(*a*).
   a. Using Ohm's law and the indicated voltage and resistance, calculate and record the current:

$$I = \frac{V}{R} = \frac{6\text{ V}}{1000\ \Omega} = \_\_\_\_\_\_\_\_\_\_\_\_\ \text{A} = \_\_\_\_\_\_\_\_\_\_\_\_\ \text{mA}$$

   b. Construct the circuit. Then measure and record the current. Remember that the circuit must be physically interrupted, as shown in Fig. 3-7(*b*), to measure current. ____________
   c. Are the measured and the calculated current about the same (within 15 percent)? If they are not, you have made an error or have faulty equipment. Check with your teacher. ____________
2. Change the 1000-Ω resistor in Fig. 3-7(*a*) to a 10,000-Ω resistor. This increases the resistance by a factor of 10 and provides the circuit shown in Fig. 3-7(*c*).
   a. According to Ohm's law, what should happen to the current?
   ____
   ____
   b. Measure and record the current with the 10,000-Ω resistor in the circuit.
   ____
   c. Is this the current you expected? ____________
3. Change the battery to a 1.5-V cell and the resistor to 1000 Ω. This will give you the circuit shown in Fig. 3-7(*d*). Notice the voltage has decreased to one-fourth the value used in Fig. 3-7(*a*).
   a. What relationship should there be between the currents in the circuits of Fig. 3-7(*a*) and (*d*)?
   ____
   ____
   b. Measure and record the current in the new circuit [Fig. 3-7(*d*)].
   ____
   c. Is this the current you would expect after decreasing the voltage from 6 V to 1.5 V? ____________

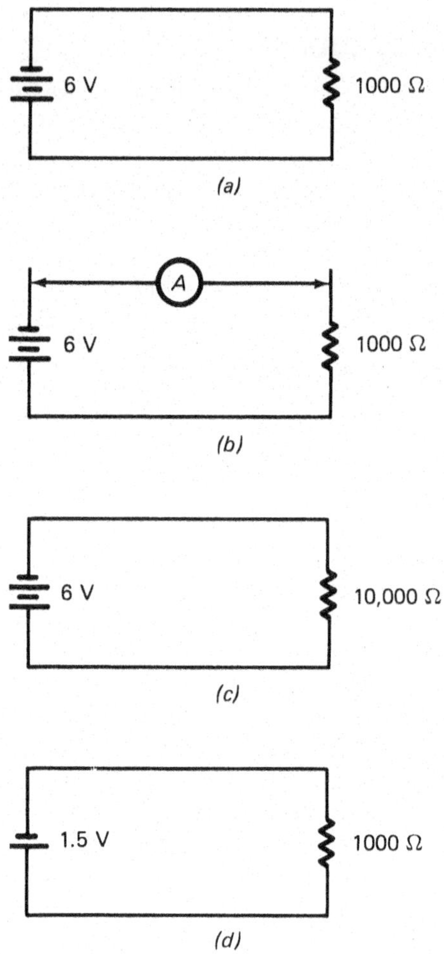

Fig. 3-7 Circuits for experiment 3-3.

NAME ______________________________ DATE ______________

3-4 LAB EXPERIMENT: TROUBLESHOOTING

PURPOSE

Electric circuits do fail. Electronic technicians need to develop troubleshooting skills. This experiment will help you do that.

MATERIALS

Qty.

1 multimeter (DMM or VOM)
1 6-V dry cell or power supply
1 no. 47 lamp
1 switch, SPST, 1A, 125 V ac
1 fuse, 0.5 A, with holder
Multisim
Multisim file for Fig. 3-8

PROCEDURE

1. The physical circuit for the schematic diagram in Fig. 3-8 has a defective component. Unless your instructor asks you to use the Multisim file for Fig. 3-8, construct the circuit with the components provided by your instructor. Using only a voltmeter, determine which component is faulty and explain how you made this determination.

 ____

 ____

2. Could there also be another defective component in the circuit? Explain.

 ____

 ____

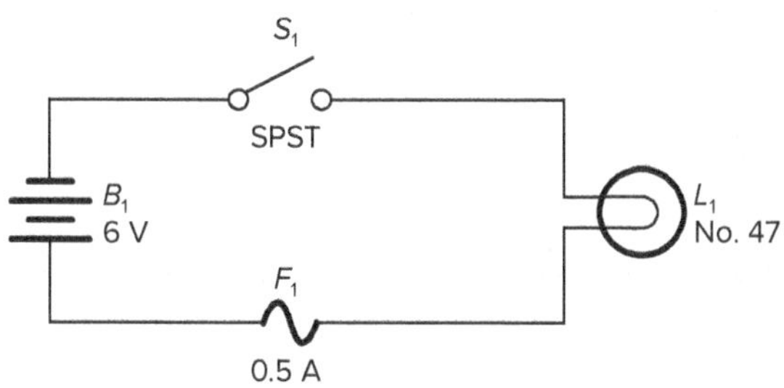

Fig. 3-8 Diagram for defective circuit.

3-5 ADVANCED PROBLEMS

3-1. How much current is flowing through a 40-Ω resistor that dissipates 360 W? ______________

3-2. A 98-Ω electric device uses 0.25 kWh of energy in 5 h. How much voltage is connected to the device? ______________

3-3. How much power is needed to force 30 C of charge through a 5-Ω resistor in 20 s? ______________

3-4. The current through a lamp is 2 A. If it requires 30 J to force 3 C through the lamp, how much power is needed? ______________

NAME ______________________________ DATE ______________

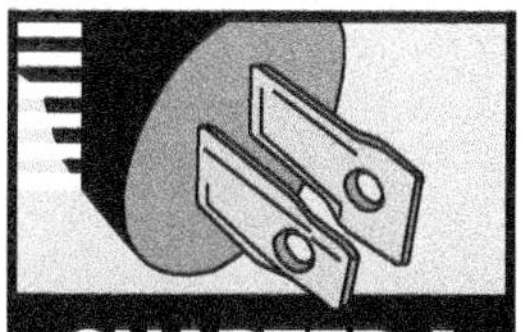

CHAPTER 4

Circuit Components

TEST: CIRCUIT COMPONENTS

For questions 1 to 9, supply the missing word or phrase needed in each statement.

1. Cells that cannot be recharged are called ______________ cells.
2. A battery has at least ______________ cells.
3. The accumulation of gas ions around an electrode in a cell is called ______________.
4. The ______________ cell has the longest shelf life of any of the primary cells.
5. A(n) ______________ cell uses sulfuric acid in its electrolyte.
6. The ______________ of a cell causes its terminal voltage to change with changes in cell current.
7. The specific gravity of a cell can be measured with a(n) ______________.
8. The specific gravity of a cell ______________ as the cell is discharged.
9. Fast-blow fuses are sometimes called ______________ fuses.

For questions 10 to 37, determine whether each statement is true or false.

10. Lead-acid batteries should be stored in a charged state.
11. The cycle life of an alkaline cell is greater than that of a nickel-cadmium cell.
12. For high-current applications, alkaline cells are better than carbon-zinc cells.
13. The carbon-zinc D cell is cheaper than the alkaline D cell.
14. For the same size cell, alkaline cells store more energy than carbon-zinc cells.
15. The alkaline cell has a less constant output voltage than the nickel-cadmium cell does.
16. Alkaline cells maintain a more constant voltage as they are discharged than mercury cells do.
17. Silver oxide cells are commonly used in high-current applications.
18. Miniature incandescent lamps have carbon filaments.
19. An incandescent lamp operated below its rated voltage has a reduced life expectancy.
20. When heated, a bimetallic strip bends.
21. The hot resistance of a lamp is greater than the cold resistance of a lamp.
22. When a neon lamp is operated on dc voltage, both electrodes glow.
23. LED circuits contain a resistor to limit the current through the LED.
24. A rheostat is often used as a potentiometer.

25. When used to describe a switch, NC is the abbreviation for "no connection." 25. ____________
26. Power resistors are usually carbon-composition resistors. 26. ____________
27. Conductive plastic is often used for potentiometers. 27. ____________
28. The resistance of a resistor is not related to the physical size of the resistor. 28. ____________
29. The power rating of a resistor is not related to the physical size of the resistor. 29. ____________
30. A switch can control more current in a 208-V circuit than it can in a 120-V circuit. 30. ____________
31. Coaxial cables are often used to connect electric motors to a power source. 31. ____________
32. A 16-gage wire has half as much resistance per foot as a 19-gage wire. 32. ____________
33. The amount of current a copper conductor can carry depends only on the cross-sectional area of the conductor. 33. ____________
34. Electric motors are usually protected with slow-blow fuses or long-time-lag circuit breakers. 34. ____________
35. Some resistors have essentially 0 Ω of resistance. 35. ____________
36. All circuit-protection devices produce an open circuit when they are tripped or blown. 36. ____________
37. An SMD has short leads for soldering into holes in a printed circuit board. 37. ____________

For questions 38 to 42, choose the letter that best completes each statement.

38. The energy storage capacity of a cell or battery is specified in 38. ____________
   a. Watthours
   b. Joule-hours
   c. Volt-hours
   d. Ampere-hours
39. Which of the following cells is a secondary cell? 39. ____________
   a. Silver oxide
   b. Carbon-zinc
   c. Nickel-cadmium
   d. Mercury
40. Which of the following switch types can perform the most complex switching? 40. ____________
   a. Rotary
   b. Toggle
   c. Rocker
   d. Push-button
41. Which of the following switch types can be either shorting or nonshorting? 41. ____________
   a. Rotary
   b. Rocker
   c. Toggle
   d. Push-button
42. The cross-sectional area of a round conductor is specified in 42. ____________
   a. Mils
   b. Circular mils
   c. Circular meters
   d. Square meters

Using the color code below, determine the resistance and tolerance indicated by the sequence of colors listed in questions 43 to 46.

Color	Value	Color	Value
Black	0	Blue	6
Brown	1 or 1 percent	Violet	7
Red	2 or 2 percent	Gray	8
Orange	3	White	9
Yellow	4	Gold	0.1 or 5 percent
Green	5	Silver	0.01 or 10 percent

43. Green, brown, silver, gold — 43. __________
44. Gray, red, green, silver — 44. __________
45. Brown, black, black, black, brown — 45. __________
46. Red, black, green, red, red — 46. __________

For questions 47 to 56, solve each problem. Be sure to include units (W, Ω, etc.) in your answers when appropriate.

47. What is the terminal voltage of a voltage source with a nominal voltage of 20 V and an internal resistance of 0.45 Ω when the source is providing 2.1 A to its load? — 47. __________

48. A 560-Ω, 5 percent resistor is at the high end of its tolerance. How much power does it dissipate when it carries 240 mA? — 48. __________

49. In an electric device, a 100-Ω, 10 percent resistor will draw 100 mA of current. What power rating should the resistor have? — 49. __________

50. How much current is drawn by a lamp rated at 12.6 V and 31.5 W? — 50. __________

51. What is the hot resistance of the lamp in question 50? — 51. __________

52. How much energy can a 24-V, 200 Ah battery provide? — 52. __________

53. Calculate the internal resistance of a 9-V battery that has a terminal voltage of 8.2 V when providing 800 mA of load current. — 53. __________

54. A resistor draws 24 mA from a 20-V power source. What power rating should the resistor have? — 54. __________

55. Calculate the minimum power dissipated by a 560-Ω, ±10 percent resistor connected to a power source with an output voltage that varies between 9 V and 11 V. — 55. __________

56. Calculate the maximum power dissipated for the conditions specified in question 55. — 56. __________

NAME ______________________________ DATE ______________

4-1 LAB EXPERIMENT: LAMP SWITCHING CIRCUIT

PURPOSE

The following experiment will provide experience in using switches and fuses. It also provides experience in building circuits from diagrams.

MATERIALS

Qty.

1 multimeter (DMM or VOM)
1 6-V dry battery or power supply
2 SPDT toggle switches (1-A, 125-V ac rating)

Qty.

1 0.5-A fuse, 32 V (or greater)
1 fuse holder
1 no. 47 lamp
1 lamp holder for no. 47 lamp

PROCEDURE

1. Using an ohmmeter, check for continuity between the center terminal and the outside terminals of one of the toggle switches. Change the switch to its other position and again check for continuity. Now check for continuity between the two outside terminals with the switch in one position and then the other. Which terminal of the switch is the pole? ______________

2. Construct the circuit shown in Fig. 4-1.

a. Throw S_1 back and forth while observing the lamp. Does the lamp turn off and on? ______________

b. Throw S_2 back and forth. Does the lamp turn off and on? ______________

c. Can S_1 control the lamp with S_2 in either position? ______________

3. While the lamp is on, remove F_1 from its holder. Removal of the fuse gives the same effect as a blown fuse. Now measure the voltage across the terminals of the fuse holder. Reinsert F_1 and again measure the voltage across its terminals. Can a voltmeter be used to indicate the condition of a fuse in a circuit? ______________ How?

____

____

4. Turn the lamp off with S_2. Now measure the voltages between the center terminal and the outside terminals of S_2. Next measure the voltages between the center terminal and the outside terminals of S_1. Can a voltmeter indicate whether or not a switch is open? ______________

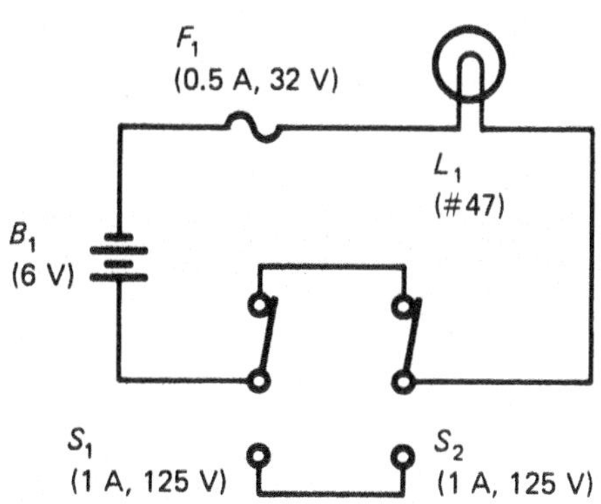

Fig. 4-1 Lamp switching circuit.

4-2 LAB EXPERIMENT: DETERMINING THE RESISTANCE OF A LAMP

PURPOSE

Completion of this experiment will acquaint you with the characteristics of incandescent lamps. The experiment will also give you additional experience in measuring current.

MATERIALS

Qty.

1 6-V dry battery or power supply
1 lamp holder
1 no. 47 lamp (6.3 V, 0.15 A)
1 VOM or DMM

PROCEDURE

1. Measure and record the resistance of the lamp. This resistance is referred to as the "cold resistance" of the lamp. It is not the same as the resistance when the lamp is operated at rated voltage. The resistance at rated voltage is called the "hot resistance." ______________

2. Determine the hot resistance by constructing the circuit of Fig. 4-2(*a*) and measuring the current (Fig. 4-2*c*) and voltage (Fig. 4-2*b*).

a. What is the measured voltage across the lamp? ______________

b. What is the measured current through the lamp? ______________

c. Now calculate and record the hot resistance with Ohm's law. ______________

d. How does this resistance compare with the resistance measured in step 1 above? ______________

3. a. Using the manufacturer's ratings (6.3 V, 0.15 A), calculate the hot resistance. ______________

b. Is this resistance closer to the resistance determined in step 1 or to that determined in step 2 above? ______________

4. a. What is the power rating of the no. 47 lamp? ______________

b. How much power was the lamp using in step 2? ______________

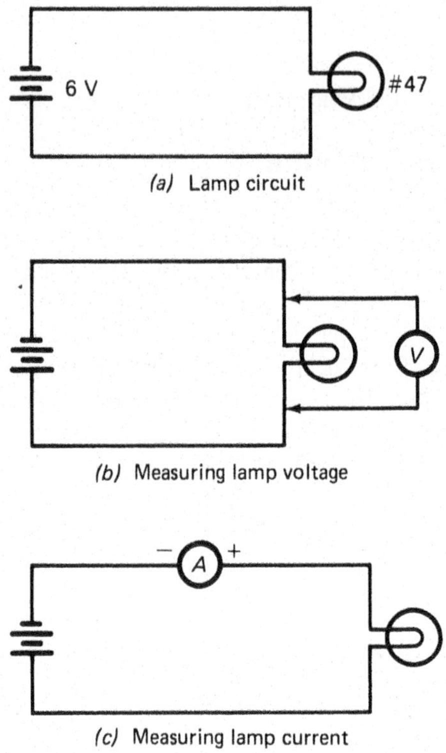

Fig. 4-2 Determining "hot resistance."

NAME ______ DATE ______

4-3 LAB EXPERIMENT: INTERNAL RESISTANCE

PURPOSE

This experiment demonstrates how the internal resistance and the terminal voltage of a cell vary as the load current varies. It also provides additional practice calculating and measuring electrical quantities.

MATERIALS

Qty.

1 multimeter (DMM or VOM)
1 1.5-V dry cell
1 100-Ω, ½-W resistor
1 47-Ω, ½-W resistor
1 22-Ω, ½-W resistor
1 10-Ω, ½-W resistor
1 4.7-Ω, ½-W resistor
1 SPST switch

PROCEDURE

1. Measure and record in Table 4-1 the terminal (NL) voltage of the dry cell when it has no resistor connected to it.
2. Construct the circuit shown in Fig. 4-3. Leave S_1 open and use the 100-Ω resistor.

TABLE 4-1 Measured and Calculated Values for Fig. 4-3

Load, Ω	$V_{measured}$, V	$I_{calculated}$, mA	V_{intR}, mV	R_{int}, Ω
NL		X	X	X
100				
47				
22				
10				
4.7				

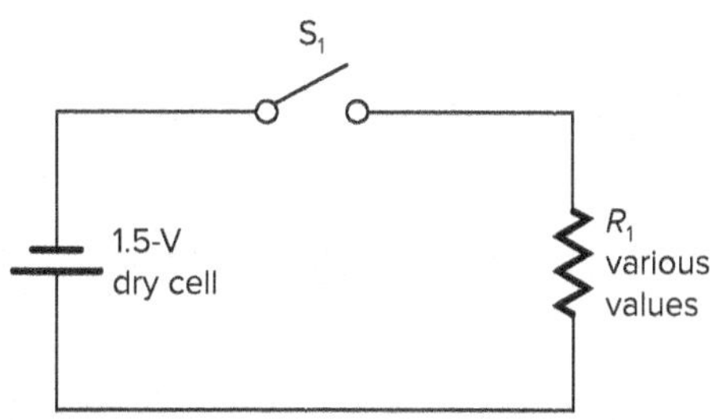

Fig. 4-3 Circuit for experiment 4-3.

3. Prepare the multimeter to measure the voltage across R_1. Close S_1 and measure the voltage across R_1. Open S_1 and record the measured voltage in Table 4-1.
4. Repeat step 3 using the other resistors listed in Table 4-1.

5. Using the measured voltages and the given value of resistance, calculate and record in Table 4-1 the current for each resistor.
6. Calculate and record in Table 4-1 the voltage dropped across the internal resistance ($V_{intR} = V_{NL} - V_{load}$).
7. Calculate and record in Table 4-1 the internal resistance for each load condition.
8. Is the decrease in internal resistance proportional to the increase in load current? ______________
9. What whould have happened if a 2.2-Ω, ½-W resistor had been used as a load?

____

10. Calculate and record the power dissipated by the 4.7-Ω load and the 47-Ω load. ______________ ______________
11. Does the 4.7-Ω resistor draw 10 times as much current as the 47-Ω resistor circuit? ______________ Why?

NAME ________________________________ DATE ______________

4-4 LAB EXPERIMENT: CHARACTERISTICS OF AN LED CIRCUIT

PURPOSE

As well as acquainting you with the operation of an LED, this experiment will provide you with more experience in building circuits and making measurements.

MATERIALS

Qty.

1 0- to 20-V dc power supply
1 DMM or VOM
1 470-Ω, ½-W resistor, ±10% resistor
1 T 1 ¾LED

• test leads and wire

PROCEDURE

1. Adjust the power-supply voltage control for minimum output, and turn off the power supply.
2. Construct the circuit shown in Fig. 4-4. Remember that an LED is polarized, so refer to Fig. 4-5 to identify the leads of the LED. The cathode lead is connected to the negative terminal of the power supply.

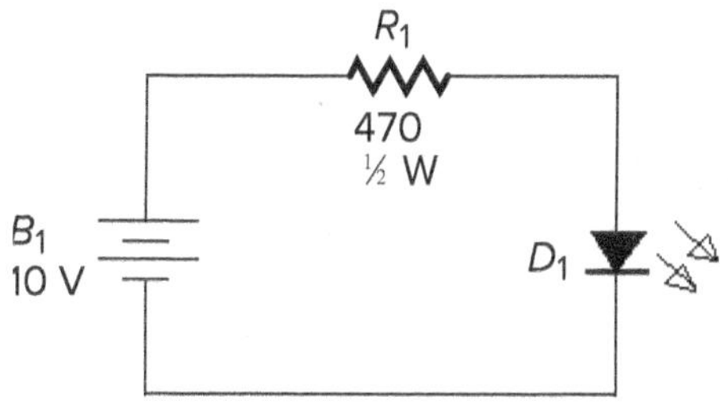

Fig. 4-4 LED characteristics.

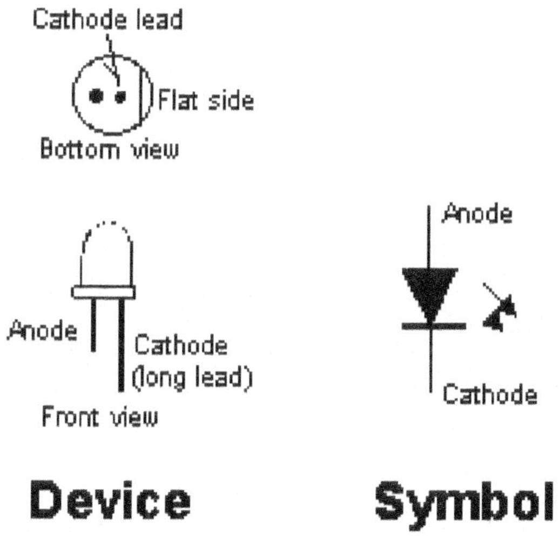

Fig. 4-5 Identifying the leads of a T 1 ¾ LED.

3. Turn on the power supply and adjust its output to 10 V.
4. Measure, and record, the voltage drop across R_1 and D_1.

V_{R_1} = ____________

V_{D_1} = ____________

5. Using the measured value of V_{R_1} and the nominal value of R_1, calculate and record the value of I_{R_1}. The value of I_{R_1} is also the value of I_{D_1}.

I_{R_1} = ____________

6. Interrupt the circuit and measure and record the current through R_1. This is also the value of I_{D_1}.

I_{R_1} = ____________

7. Calculate and record the resistance of the resistor using the measured value of V_{R_1} and I_{R_1}.

R_1 = ____________

8. Does the resistance of R_1 appear to be within its tolerance limits?

____

9. Reduce the source voltage to the value where you can just discern that the LED is glowing. Now measure and record the values of I_{R_1} and V_{D_1}.

I_{R_1} = ____________

V_{D_1} = ____________

10. Compare the values recorded in step 9 with those recorded in steps 4 and 6. What happens to the values of I_{D_1} and V_{D_1} as the output of the LED gets dimmer? ____

____

4-5 LAB EXERCISE: RESISTOR COLOR CODE, TOLERANCE, AND STANDARD VALUES

1. Refer to Fig. 4-23 in chapter 4 of your textbook. List the values of resistors between 400 and 600 Ω that are available in 5 and 10 percent tolerances.

   5% ________ ________ ________ ________

   10% ________ ________ ________ ________

2. Notice that there are fewer values available in 10 percent resistors than in 5 percent resistors. For example, in a 5 percent resistor you can obtain values of 470, 510, and 560 Ω. Only the 470-Ω and 560-Ω values are obtainable in 10 percent tolerance. This is because the minimum value of a ±10 percent, 560-Ω resistor is less than 510 Ω. And the maximum value of a ±10 percent, 470-Ω resistor is greater than 510 Ω.
3. Standard 680- and 750-Ω resistors are available in 5 percent tolerance.
   a. Without referring to Fig. 4-23, determine whether or not a 715-Ω, ±5 percent resistor is available. Is it? ____________ Why?

   ____

   ____

   b. Would it be available in a 10 percent resistor? ____________
   c. Would it be available in a 1 percent resistor? ____________
4. Obtain ten different color-coded resistors from your instructor. The resistors can have any value in the range extending from 1 Ω to 2.2 MΩ. For each resistor, record the color of each band in the spaces provided below. Next, determine the resistors color-coded value and record it under the column labeled "Coded Value". The coded value should include both resistance and tolerance values. Next, measure the value of each resistor and record your answer under the column labeled "Measured Value".

	1st Band	2nd Band	3rd Band	4th Band	Coded Value	Measured Value
R_1	______	______	______	______	______	______
R_2	______	______	______	______	______	______
R_3	______	______	______	______	______	______
R_4	______	______	______	______	______	______
R_5	______	______	______	______	______	______
R_6	______	______	______	______	______	______
R_7	______	______	______	______	______	______
R_8	______	______	______	______	______	______
R_9	______	______	______	______	______	______
R_{10}	______	______	______	______	______	______

When comparing the measured and coded values, are any resistors out of tolerance? If so, identify which ones. ______________________

____

4-6 LAB EXERCISE: USING WIRE CHARTS AND TABLES

This exercise requires the use of wire tables. These are available in electrical handbooks and manufacturers' catalogs. They can also be located by searching the Internet.

1. An electric saw is to be operated on a 100-ft extension cord. The saw requires 10 A of current.
   a. What size conductors will be required? ______________
   b. Why does a 125-ft cord require a larger conductor than a 100-ft cord to carry 10 A of current?

   ____

   ____

2. What is the resistance of 140 ft of 16-gage solid copper wire at 20°C?

   ____

3. In residential wiring, the electrical code permits a 14-gage copper conductor to carry 15 A under certain conditions. What is the rating of this conductor in circular mils per ampere? ______________

NAME ______________________________ DATE ______________

CHAPTER 5

Multiple–Load Circuits

TEST: MULTIPLE-LOAD CIRCUITS

For questions 1 to 8, supply the missing word or phrase in each statement.

1. All the voltages in a(n) ______________ circuit are equal. **1.** ______________

2. All the currents in a(n) ______________ circuit are the same. **2.** ______________

3. The reciprocal formula is used to calculate total resistance in a(n) ______________ circuit. **3.** ______________

4. Adding another series resistor to a series circuit will ______________ the total conductance. **4.** ______________

5. In a parallel circuit, the total resistance must be ______________ than the lowest individual resistance. **5.** ______________

6. Ammeters have a very ______________ internal resistance. **6.** ______________

7. The total power of a parallel circuit ______________ when one load opens. **7.** ______________

8. The ______________ resistance drops the most voltage in a series circuit. **8.** ______________

For questions 9 to 20, determine whether each statement is true or false.

9. In a parallel circuit, one end of a resistor must be disconnected from the circuit to measure its resistance. **9.** ______________

10. The total current in a series circuit can be measured by inserting an ammeter in series with any load in the circuit. **10.** ______________

11. A deficiency of electrons always exists at the positive end of a resistor. **11.** ______________

12. The total resistance of a series circuit is always less than the lowest resistance. **12.** ______________

13. The voltage measured across an open load is approximately equal to the source voltage. **13.** ______________

14. In a parallel circuit, the lowest resistance uses the least power. **14.** ______________

15. When one load is removed from a parallel circuit, the total resistance decreases. **15.** ______________

16. Maximum power transfer in a circuit leads to maximum efficiency. **16.** ______________

17. For maximum power transfer, the load resistance should be not more than half as high as the source resistance. **17.** ______________

18. A low resistance has more conductance than a high resistance. **18.** ______________

19. When any one of the resistors in a parallel circuit is removed, the conductance of the circuit increases. **19.** ______________

20. Three 33-Ω resistors connected in parallel provide 99 Ω of total resistance. **20.** ______________

For questions 21 to 31, choose the letter that best completes each statement.

21. A series-parallel circuit is classified as **21.** __________
a. Single-load circuit
b. Single-path circuit
c. Multiple-load circuit
d. Multiple-path circuit

22. The formula $P_T = P_{R_1} + P_{R_2} + P_{R_3}$ is used with **22.** __________
a. Series circuits
b. Parallel circuits
c. Series-parallel circuits
d. All the above circuits

23. The formula $I_T = I_{R_1} + I_{R_2} + I_{R_3}$ is used with **23.** __________
a. Series circuits
b. Parallel circuits
c. Series-parallel circuits
d. All the above circuits

24. The formula $R_T = R_1 + R_2 + R_3$ is used with **24.** __________
a. Series circuits
b. Parallel circuits
c. Series-parallel circuits
d. All the above circuits

25. The highest resistor dominates the circuit if the circuit is **25.** __________
a. Series
b. Parallel
c. Series-parallel
d. Any of the above

26. The total resistance of 68-, 82-, and 100-Ω resistors connected in parallel will be **26.** __________
a. Less than 68 Ω
b. Between 68 and 82 Ω
c. Between 82 and 100 Ω
d. Greater than 100 Ω

27. When estimating the total resistance of the circuit in Fig. 5-1, which resistor would you omit from your calculations? **27.** __________
a. R_1 **c.** R_3
b. R_2 **d.** R_4

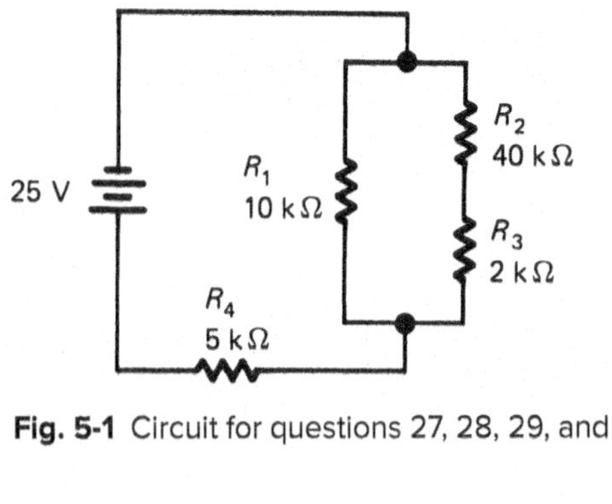

Fig. 5-1 Circuit for questions 27, 28, 29, and 36.

28. In Fig. 5-1, increasing R_1 to 15 kΩ will **28.** __________
a. Increase I_T **c.** Increase P_T
b. Increase R_T **d.** Increase V_T

29. In Fig. 5-1, decreasing R_2 to 10 kΩ will **29.** __________
a. Decrease V_{R_4} **c.** Decrease V_{R_1}
b. Decrease V_{R_3} **d.** Decrease V_T

NAME ______________________________ DATE ______________

30. The current in an unloaded voltage divider is called the

a. Residual source current
c. Bleeder current
b. Divider current
d. Lost current

30. ______________

31. The zener diode should be operated with

a. No bias
b. Reverse bias
c. Forward bias
d. Sideways bias

31. ______________

Solve the following problems and record your answers. Be sure to specify units as well as numbers.

32. Find the values for V_{R_1}, R_2, R_3, P_T, and R_T for Fig. 5-2.

32. ______________

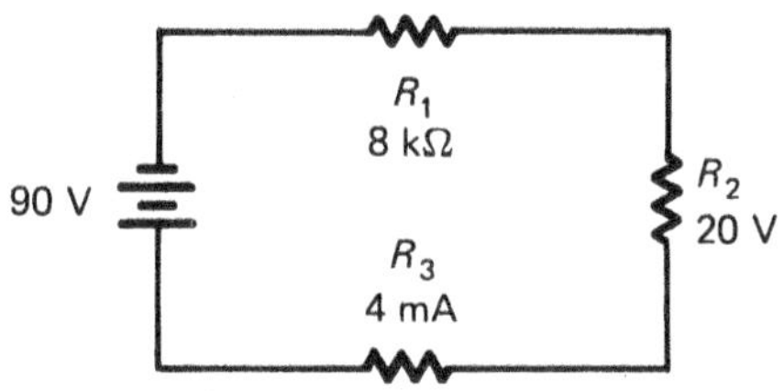

Fig. 5-2 Circuit for question 32.

33. Find the values for I_{R_1}, I_{R_2}, R_2, P_T, and P_{R_1} for Fig. 5-3.

33. ______________

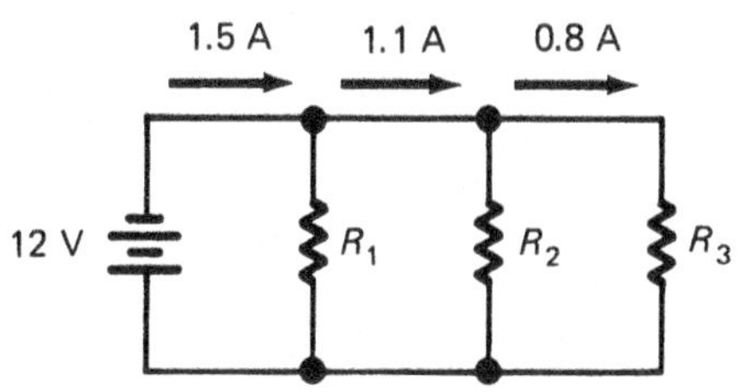

Fig. 5-3 Circuit for question 33.

34. Three resistors—22, 44, and 66 Ω—are parallel-connected to a 6-V battery. Determine the values for R_T, I_T, P_T, and G_T.

34. ______________

35. For the circuit of Fig. 5-4, find the values for V_{R_1}, V_{R_3}, I_{R_1}, R_1, and R_T. **35.** __________

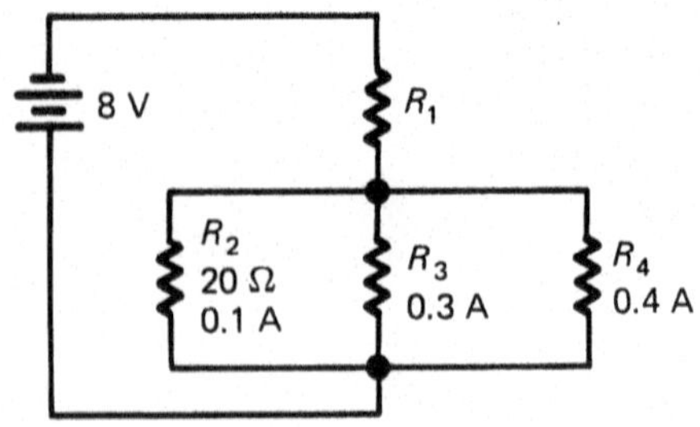

Fig. 5-4 Circuit for question 35.

36. Determine R_T, I_T, and P_T for Fig. 5-1. **36.** __________

37. Which, if any, resistors in Fig. 5-5 are directly in parallel? **37.** __________

38. Which, if any, resistors in Fig. 5-5 are directly in series? **38.** __________

39. For Fig. 5-5 determine V_{R_1}, V_{R_3}, I_{R_4}, I_T, and R_T. **39.** __________

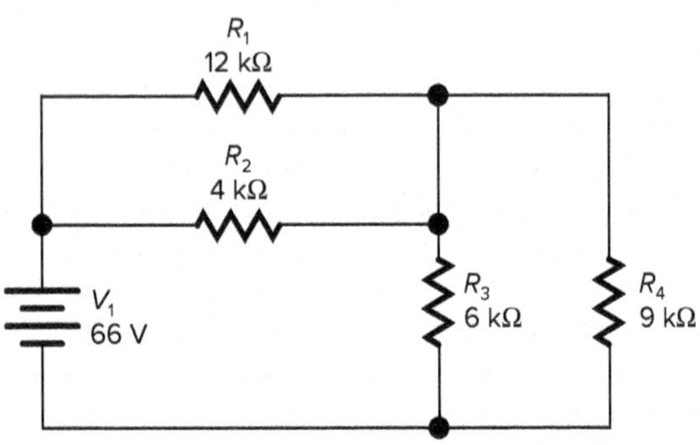

Fig. 5-5 Circuit for questions 37, 38, and 39.

NAME ______________________________ DATE ______________

5-1 LAB EXPERIMENT: SERIES CIRCUITS—GENERAL PROPERTIES

PURPOSE

This experiment is designed to illustrate the major characteristics of a series circuit. It will also provide experience in constructing and testing series circuits.

MATERIALS

Qty.		Qty.	
1	6-V dry battery (or 6-V dc power supply)	1	4.7-Ω, ½-W resistor
2	no. 47 lamps	1	33-Ω, ½-W resistor
2	lamp holders for no. 47 lamps	1	VOM or DMM
		•	test leads and wire

PROCEDURE

Note: If using a battery, operate the lamps for the shortest time possible to conserve the battery.

1. Connect the circuit shown in Fig. 5-6(*a*).

a. Do the lamps glow with equal brilliancy? ______________ Why?

____

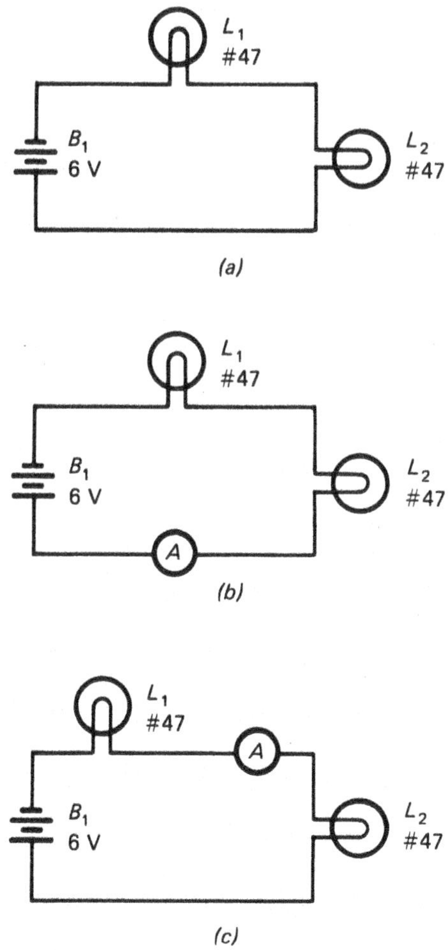

Fig. 5-6 Series lamps.

b. Does either lamp glow with normal brilliancy? ______________ Why?

____

c. Should $V_{L_1} + V_{L_2}$ equal the battery voltage? ______________

d. Measure and record the voltage across each lamp and across the battery.
V_{L_1} ______________ V_{L_2} ______________ V_{B_1} ______________

e. Does $V_{L_1} + V_{L_2}$ equal the battery voltage? ______________

f. Interrupt the circuit between the positive terminal of the power source and L_2. Insert the DMM or VOM (current function, 500-mA range) as shown in Fig. 5-6(*b*). Measure and record the current at this location in the circuit. ______________

g. Should the current be the same if measured between L_1 and L_2, as indicated in Fig. 5-6(*c*)? Check your answer by reconnecting the DMM or VOM according to the diagram of Fig. 5-6(*c*). ______________

h. Is the current measured in Fig. 5-6(*b*) and (*c*) the same?
____

2. Remove the DMM from the circuit and reconnect the circuit as in Fig. 5-6(*a*).

a. Will L_2 become more brilliant or less brilliant if L_1 is shorted out? ______________ Why?

____

____

b. Short out L_1 by connecting a jumper lead from one terminal of the socket to the other terminal of the socket. Does the brilliancy of L_2 change as you expected? ______________

c. Remove the jumper from the socket of L_1. If L_2 is shorted out, what will the effect be on the brilliancy of L_1? Check your conclusion.

____

____

3. Remove any shorting jumpers so that the circuit is back to what is shown in Fig. 5-6(*a*).

a. Now remove L_1 from its socket. Removing L_1 has the same effect as L_1 burning out. What happened to the light output of L_2?
______________ Why?

____

____

b. How much voltage would you expect across L_2 when L_1 is removed? ______________ Why?

____

____

c. Verify your answer by measuring and recording the voltage across L_2.
____

d. How much voltage would you expect across the terminals of the socket from which L_1 was removed? ______________ Why?

____

____

e. Verify your answer by measuring and recording the voltage across the terminals of the socket. ______________

f. Do your results support the statement that "the voltage across an open load equals the source voltage in a series circuit"? ______________

4. Construct the circuit shown in Fig. 5-7, using a 4.7-Ω resistor for R_1. Note the brilliancy of the lamp.

a. How does it compare with the brilliancy it had when the circuit of Fig. 5-6(*a*) was used?

____

____

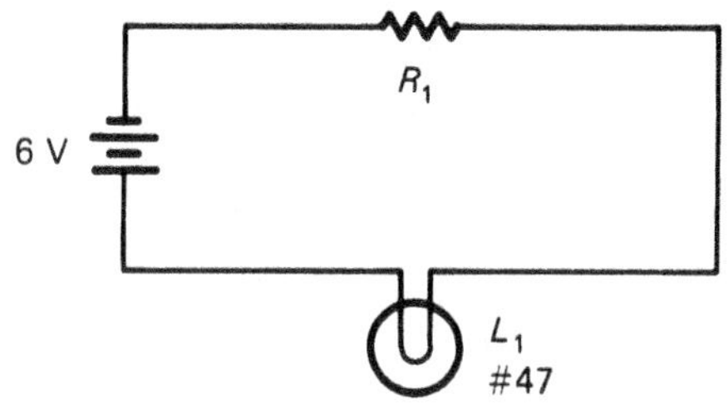

Fig. 5-7 Controlling lamp current.

b. Which do you think is dropping more voltage, R_1 or L_1?
______________ Why?
____

____

c. To test your conclusions, measure the voltages across R_1 and L_1. Which has more resistance, R_1 or L_1? ______________

d. If the 4.7-Ω resistor at R_1 is replaced with a 33-Ω resistor, will L_1 get brighter or dimmer? ______________

e. Test your answer by changing R_1 to 33 Ω. Is the voltage across R_1 or L_1 greater now? ______________

f. Is the resistance of R_1 greater or less than the resistance of L_1?
____

g. Could a 100-Ω rheostat at R_1 control the output of L_1 from no light to full brilliancy? ______________

5-2 LAB EXPERIMENT: PARALLEL CIRCUITS—GENERAL PROPERTIES

PURPOSE

This experiment illustrates the major characteristics of parallel circuits. Additional practice in constructing circuits and making voltage, current and resistance measurements is also provided.

MATERIALS

Qty.

1 6-V dry battery (or 6-V dc power supply)
2 no. 47 lamps
2 lamp holders for no. 47 lamps

Qty.

1 220-Ω, ½-W resistor
1 VOM or DMM
• test leads and wire

PROCEDURE

1. Construct the circuit in Fig. 5-8(*a*).

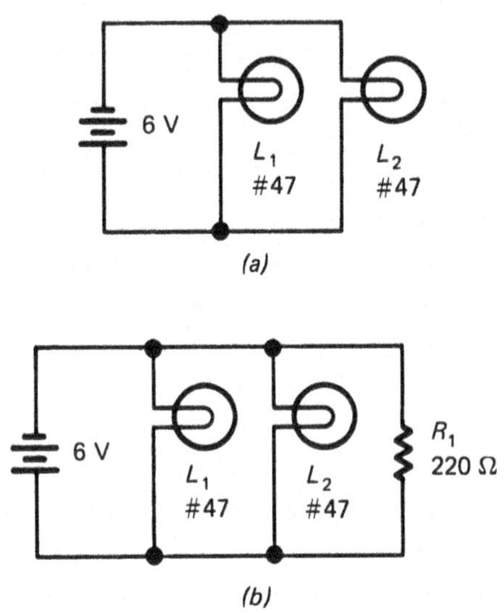

Fig. 5-8 Parallel loads.

a. Measure the voltage across the terminals of the battery and across the terminals of each lamp socket. Are all measured voltages the same? ______________

b. Remove L_1 from its socket and measure the voltage across the terminals of the L_1 socket. Did the voltage at the terminals of the L_1 socket change when L_1 was removed? ______________ Why?

____

____

c. Did the brilliancy of L_2 change when L_1 opened (was removed)? ______________

d. Put L_1 back in its socket and remove L_2. Did removal of L_2 affect the brilliancy of L_1? ______________

e. Reinstall L_2 in its socket. Add a 220-Ω resistor in parallel, as shown in Fig. 5-8(*b*). Did the addition of the resistor change the brilliancy of L_1 and L_2? ______________

f. Do your data indicate that each branch of a parallel circuit is independent of the other branches? ______________

2. Set the DMM or VOM to the current function and the 500-mA range. Replace the conductor between the negative terminal of the battery and the terminal of L_1 with the DMM or VOM. This will measure the total (source) current in Fig. 5-8(*b*).
   a. Remove L_1. Did the total current increase or decrease?
   ____
   b. Remove L_2. Did the total current increase or decrease?
   ____
   c. Which draws more current from the voltage source, the 220-Ω resistor or L_2? ______________
   d. Does the sum of the branch currents equal the total current?
   ____

5-3 LAB EXPERIMENT: SERIES-PARALLEL CIRCUITS—GENERAL PROPERTIES

PURPOSE

The following experiment provides experiences in constructing more complex circuits. It will allow you to determine some of the major properties of series-parallel circuits. This experiment provides clear visual evidence of the interdependency of these circuits.

MATERIALS

Qty.		Qty.	
1	6-V dry battery (or 6-V dc power supply)	4	lamp holders for no. 47 lamps
4	no. 47 lamps	•	jumper leads and wire

PROCEDURE

1. Study the circuit in Fig. 5-9 and predict the answers to the following questions:

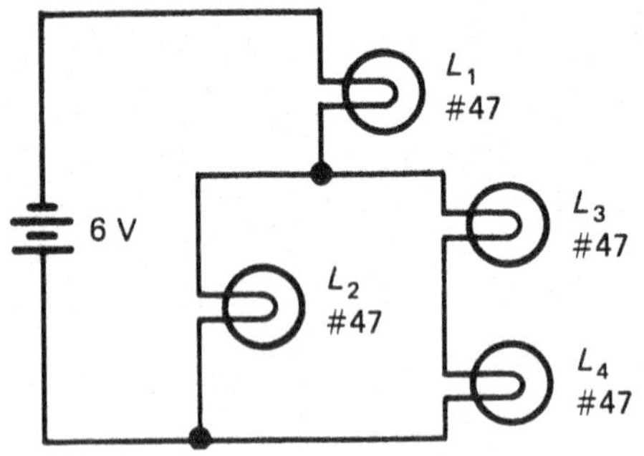

Fig. 5-9 Series-parallel lamps.

   a. Which lamp(s) will be most brilliant? ______________

   b. Which lamp(s) will be least brilliant? ______________

   c. Now construct the circuit of Fig. 5-9 and check your answers to questions a and b. Why is lamp L_1 more brilliant than lamp L_2?

   ____

   ____

   d. Why is L_2 more brilliant than either L_3 or L_4?

   ____

   ____

   e. Why are the brilliancies of L_3 and L_4 equal?

   ____

   ____

2. Again study Fig. 5-9 and predict what will happen to the brilliancy of the other lamps when L_1 is shorted out. Then short out L_1 with a jumper lead and see what happens.

   a. Did L_2 get brighter or dimmer? ______________

   b. Did L_3 and L_4 get brighter or dimmer? ______________

   c. Explain why the brilliancy of L_2 changed as it did when L_1 was shorted.

   ____

   ____

   d. Remove the jumper lead (short) from L_1. Now predict the effect on the brilliancy of the other lamps when L_2 is shorted. Are you sure of your predictions? If so, short out L_2 and record your results. Did L_1 get brighter or dimmer? ______________

e. Did L_3 and L_4 get brighter or dimmer? ______________

f. Explain why L_1 and L_3 changed as they did when L_2 was shorted. Remove the short from L_2.

____

____

g. If L_3 is shorted, will L_2 get brighter or dimmer? ______________

h. If L_4 is shorted, will L_2 get brighter or dimmer? Check your answers by shorting out L_3 and then L_4. ______________

i. Why does L_1 get brighter when either L_3 or L_4 is shorted?

____

____

j. Why does L_3 get brighter when L_4 is shorted?

____

____

5-4 LAB EXERCISE: POLARITIES AND MEASUREMENTS

PURPOSE

This exercise will provide additional experience in determining the polarity of voltage drops in multiload circuits. It will also illustrate current and voltage measurements in these circuits.

PROCEDURE

1. Refer to Fig. 5-10(*a*). The polarity of the voltage across R_1 is already marked on the diagram.

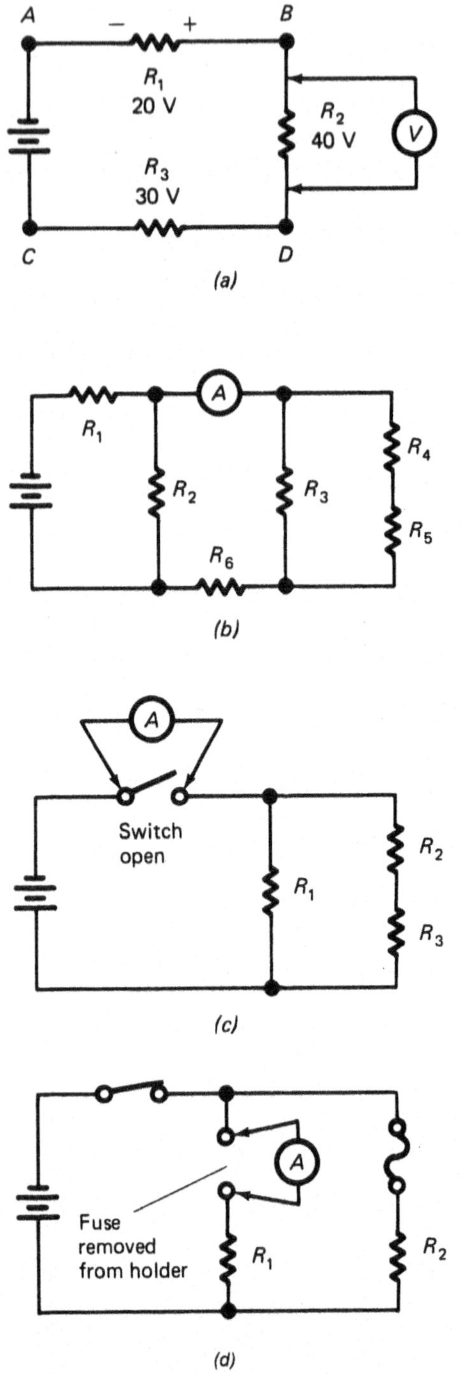

Fig. 5-10 Polarities and measurements.

a. Mark the polarity of the voltages across R_2 and R_3. Also indicate the polarity of the voltmeter. In Fig. 5-10(*a*), the voltage at point *B* is 20 V positive with respect to point *A*.
b. What is the voltage at point *C* with respect to point *B*? ______
c. What is the voltage at point *C* with respect to point *D*? ______

2. a. In Fig. 5-10(*b*), what current is being measured by the ammeter? ______
   b. Which lead (– or +) of the ammeter should be connected to the junction of R_1 and R_2 in Fig. 5-10(*b*)? ______
   c. Will the current in R_6 flow from left to right or from right to left? ______
3. Usually when current is measured, the circuit must be physically interrupted to insert the ammeter in series. Sometimes the desired current can be measured by connecting the ammeter across an open switch or fuse.
   a. What current is being measured in Fig. 5-10(*c*)? ______
   b. Should the negative or the positive lead of the ammeter be connected to the battery side of the switch? ______
   c. What current is being measured in Fig. 5-10(*d*)? ______
   d. Which lead of the ammeter should be connected to the R_1 side of the fuse holder? ______
4. In the space next to Fig. 5-10, redraw the circuit of Fig. 5-10(*b*) and show where an ammeter would be inserted to measure I_{R_3}.

5-5 LAB EXPERIMENT: SERIES CIRCUITS—CURRENT, VOLTAGE, AND RESISTANCE

PURPOSE

This experiment will provide experience in measuring and calculating voltage, current and resistance values in series circuits. It will show that measured and calculated values are not exactly equal because of component tolerances.

MATERIALS

Qty.		Qty.		
1	20-V dc power supply	1	1-kΩ, 1-W resistor	Multisim
2	10-kΩ, ½-W resistors	1	VOM or DMM	
1	20-kΩ, ½-W resistor	•	test leads and wires	

PROCEDURE

1. **a.** Calculate and record the total resistance of the circuit in Fig. 5-11(*a*). ______________
   b. Connect the two 10-kΩ resistors in series (*do not* connect them to the power supply) and measure and record their total resistance. ______________
   c. Now connect the circuit of Fig. 5-11(*a*) and measure and record the total current. ______________
   d. Calculate and record the total current. ______________
   e. In a series circuit, do equal resistors drop the same amount of voltage? ______________
   f. Measure and record the voltages across R_1 and R_2 in Fig. 5-11(*a*). V_{R_1} ______________ V_{R_2} ______________
   g. Are these two voltages equal (within 10 percent)? ______________
   h. Now calculate and record the voltage across R_1 and across R_2. V_{R_1} ______________ V_{R_2} ______________
   i. How much power is required from the battery in Fig. 5-11(*a*)? ______________
   j. Will R_1 and R_2 each dissipate the same amount of power? ______________

TABLE 5-1 Electrical Quantities for Fig. 5-11(*b*)

Quantity	Measured Value		Calculated Value
	Multisim	Hardware	
R_T			
I_T			
V_{R_1}			
V_{R_2}			

2. **a.** If your instructor asks you to perform this experiment using Multisim, record the measured values in the Multisim column of Table 5-1. When you use Multisim, the measured and calculated columns should agree within normal round-off error. Compare them as you do the experiment.

   If you are using physical (real) components, record measured values in the Hardware column of Table 5-1. Keep comparing your measured and calculated values. If they disagree by more than 10 percent, you have probably made an error in either your calculations or your measurements. Repeat either or both, as necessary.

NAME DATE

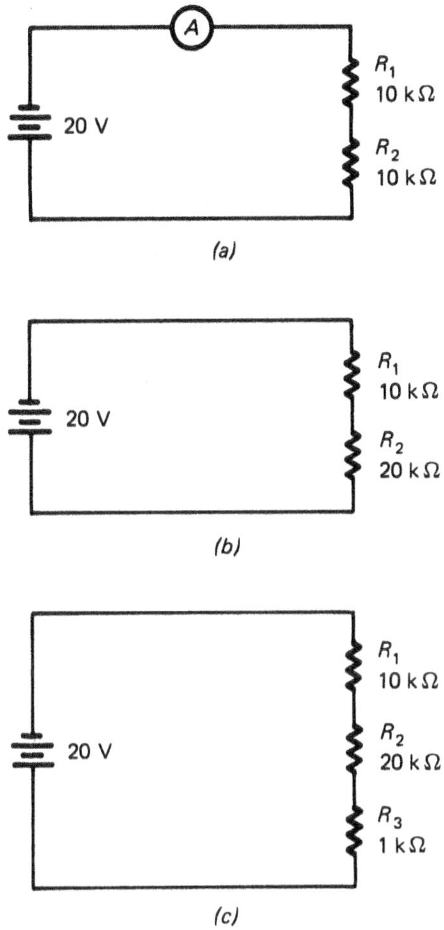

Fig. 5-11 Circuits for experiment 5-5.

Using the circuit of Fig. 5-11(*b*), make the measurements and calculations needed to fill in Table 5-1. In completing Table 5-1, calculate one quantity and then measure the same quantity. If the two agree, then do the next calculation and measurement. By following this procedure, you will discover and correct mistakes that might otherwise affect the rest of the experiment.

b. Is V_{R_2} twice as large as V_{R_1}? ______________

c. Would the voltage across a 30-kΩ resistor be three times as large as the voltage across a 10-kΩ resistor? ______________

d. If a 5-kΩ resistor drops 10 V, how much voltage will be dropped across a 20-kΩ resistor that is in series with the 5-kΩ resistor? Notice that the voltage drops in a series circuit are in direct proportion to the individual resistances. Mathematically, this statement can be written

$$\frac{R_1}{R_2} = \frac{V_{R_1}}{V_{R_2}}$$

When rearranged to solve for V_{R_2}, this relationship becomes

$$V_{R_2} = \frac{V_{R_1} R_2}{R_1}$$

Check your answer by using this equation (the 5-kΩ resistor is R_1).

e. The power in a series circuit also distributes in direct proportion to the individual resistances. Check this statement by calculating the power of R_1 and R_2 in Fig. 5-11(*b*).

P_{R_1} ______________ P_{R_2} ______________

f. Is P_{R_2} twice as large as P_{R_1}? ______________

3. Construct the circuit of Fig. 5-11(*c*).

a. Would you expect the total current of this circuit to be about the same as the total current of Fig. 5-11(*b*)? ______________ Why?

____

b. Measure the total current. Is it 3 to 4 percent lower than in Fig. 5-11(*b*)?

____

c. Would you expect V_{R_2} of Fig. 5-11(*b*) and (*c*) to be about the same? Measure V_{R_2} in Fig. 5-11(*c*) to check your answer. ______________

d. Using the formula in step 2d and the measured value of V_{R_2}, predict V_{R_3}. Now measure V_{R_3} to check your answer.

Predicted V_{R_3} ______________

e. Would shorting out R_3 affect the total power by less than or more than 5 percent? ______________

f. Would the total power increase or decrease? ______________

NAME ______________________________ DATE ______________

5-6 LAB EXPERIMENT: PARALLEL CIRCUITS—CURRENT, VOLTAGE, AND RESISTANCE

PURPOSE

One of the objects of this experiment is to verify Kirchhoff's current law. The other objective is to provide experience in measuring and calculating voltage, current and resistance values in parallel circuits.

MATERIALS

Qty.		Qty.		
1	20-V dc power supply	1	1-kΩ, 1-W resistor	Multisim
2	10-kΩ, ½-W resistors	1	VOM or DMM	
1	20-kΩ, ½-W resistor	•	test leads and wires	

PROCEDURE

If your instructor asks you to perform this experiment using Multisim, record the measured values in the Multisim column of Tables 5-2 and 5-3. When you use Multisim, the measured and calculated columns should agree within normal round-off error. Compare them as you do the experiment.

TABLE 5-2 Electrical Quantities for Fig. 5-12(*a*)

Quantity	Measured Value		Calculated Value
	Multisim	Hardware	
R_T			
I_{R_1}			
I_{R_2}			
I_T			
V_{R_1}			
V_{R_2}			

If you are using physical (real) components, record measured values in the Hardware column of Tables 5-2 and 5-3.

1. Refer to Fig. 5-12(*a*). Would you expect
   a. R_1 to be greater than R_T? ______________
   b. I_{R_1} to be greater than I_T? ______________
   c. I_{R_1} to be greater than I_{R_2}? ______________
   d. Connect two 10-kΩ resistors in parallel and measure the equivalent (total) resistance. Enter the measured value of R_T in Table 5-2. Was your answer to question correct? Check your response to questions b and c by connecting the circuit and finishing Table 5-2. In completing Table 5-2, make a calculation and then make the corresponding measurement before proceeding to the next calculation.
   e. Do your measured values of current agree with Kirchhoff's current law? ______________
   f. Compute the values of P_{R_1}, P_{R_2}, and P_T by using the measured data in Table 5-2.
   P_{R_1} ______________ P_{R_2} ______________ P_T ______________
   g. Do your measured data show that $P_T = P_{R_1} + P_{R_2}$? ______________

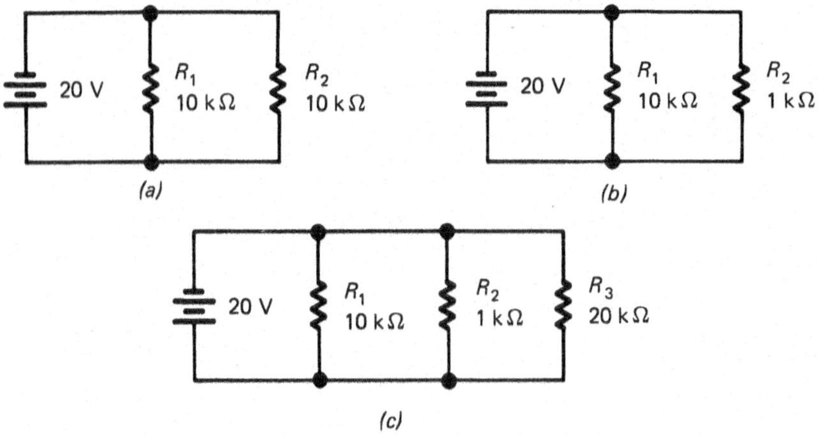

Fig. 5-12 Circuits for experiment 5-6.

2. a. Construct the circuit shown in Fig. 5-12(*b*) and complete Table 5-3. Remember to measure the resistance before power is applied to the circuit.

TABLE 5-3 Electrical Quantities for Fig. 5-12(*b*)

Quantity	Measured Value		Calculated Value
	Multisim	Hardware	
R_T			
I_{R_1}			
I_{R_2}			
I_T			

b. Does changing the value of R_2 from 10 to 1 kΩ affect the current in R_1?

____

c. Does the sum of the branch currents ($I_{R_1} + I_{R_2}$) equal the total current I_T?

____

d. Which resistor, R_1 or R_2, uses more power? _______________

3. Construct the circuit shown in Fig. 5-12(*c*).

a. Measure and record R_T, I_T, and I_{R_3}.

R_T _______________ I_T _______________ I_{R_3} _______________

b. Would the measured values of I_{R_1} and I_{R_2} be the same as they were for the circuit of Fig. 5-12(*b*)? _______________

c. Is I_{R_3} equal to the difference between the total current of Fig. 5-12(*c*) and the total current of Fig. 5-12(*b*)? _______________

d. How much current flows in the conductors that connect R_1 to R_2 in Fig. 5-12(*c*)? Check your answers by measuring the current in the conductors. _______________

4. Answer the following questions without referring to your measured or calculated data. Then, check your answers by referring to the data of steps 1 to 3.

a. Which circuit (*a*, *b*, or *c*) in Fig. 5-12 will

(1) have the least resistance? _______________

(2) draw the most current? _______________

(3) consume the most power? _______________

b. Will the total resistance of Fig. 5-12(*b*) be less than or more than 1 kΩ?

c. Will the total resistance of Fig. 5-12(*c*) be less than or more than 1 kΩ?

d. Which resistor in Fig. 5-12(*c*) will

(1) draw the least current? _______________

(2) dissipate the most power? _______________

(3) dominate the circuit? _______________

NAME ______________________________ DATE ______________

5-7 LAB EXPERIMENT: SERIES-PARALLEL CIRCUITS—CURRENT, VOLTAGE, AND RESISTANCE

PURPOSE

This experiment will provide experience in making measurements in multiple-load circuits. It will also give additional practice in computing voltage, current and resistance values in series-parallel circuits.

MATERIALS

Qty.		Qty.		
1	20-V dc power supply	1	1-kΩ, 1-W resistor	Multisim
2	10-kΩ, ½-W resistors	1	VOM or DMM	
1	20-kΩ, ½-W resistor	•	test leads and wires	

PROCEDURE

If your instructor asks you to perform this experiment using Multisim, record the measured values in the Multisim column of Table 5-4. When you use Multisim, the measured and calculated columns should agree within normal round-off error. Compare them as you do the experiment.

If you are using physical (real) components, record the measured values in the Hardware column.

Look at Fig. 5-13, and visualize how it could be reduced to an equivalent circuit containing two parallel resistors. Now, using this "visualized circuit," estimate the total resistance.

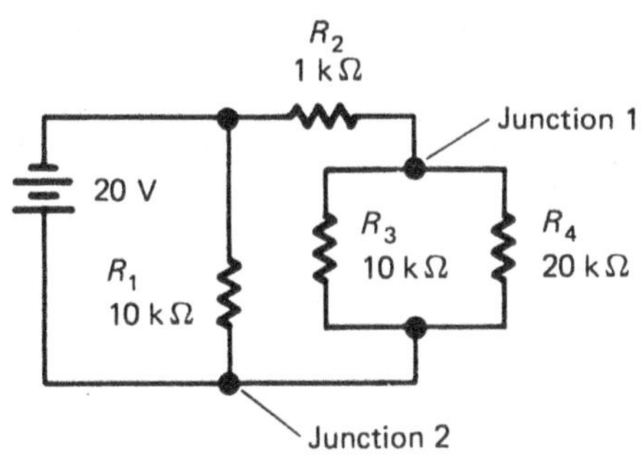

Fig. 5-13 Circuit for experiment 5-7.

1. Will R_T be greater than or less than 10 kΩ? ______________
2. Will the current through R_1 be equal to, less than, or more than the current through R_2? ______________
3. Construct the circuit of Fig. 5-13 and complete Table 5-4. Then check your answers to steps 1 and 2.
4. Using measured values of R_T and I_T, compute and record P_T. ______________
5. Use your calculated values of R_T and I_T and again compute and record P_T. ______________
6. Are your answers to steps 4 and 5 exactly the same? ______________
7. If they are not, why do you think they are different?

____

____

TABLE 5-4 Electrical Quantities for Fig. 5-13

Quantity	Measured Value		Calculated Value
	Multisim	Hardware	
R_T			
V_{R_1}			
V_{R_2}			
V_{R_4}			
I_{R_1}			
I_{R_2}			
I_{R_3}			
I_{R_4}			
I_T			

8. What is the potential difference (voltage) between junction 1 and junction 2 in Fig. 5-13? ______________
9. With respect to junction 2, is junction 1 negative or positive? Check your answers to this step and step 8 by connecting the voltmeter between the two junctions. ______________
10. Does the current entering junction 1 equal the current leaving junction 1? ______________
11. Would removing R_1 from the circuit change the value of I_{R_3}? Check this by temporarily removing R_1 and measuring I_{R_3}. ______________
12. Would removing R_4 change the value of I_{R_1}? Again, check your answer by removing R_4 while measuring I_{R_1}. ______________
13. Does removing R_4 change I_{R_2}? ______________
14. If a 100-Ω resistor were added in parallel with R_4, what would happen to the voltage across R_2? ______________

NAME ______________________________ DATE ______________

5-8 LAB EXPERIMENT: ZENER VOLTAGE DIVIDER AND REGULATOR

REGULATOR

PURPOSE

More experience in circuit construction, measuring voltage and currents, and calculating circuit values can be gained by completing this experiment.

MATERIALS

Qty.

1 20-V power supply
2 470-Ω, ½-W resistors
1 1-kΩ, ½-W, ±5% resistor
1 2.2-kΩ, ½-W, ±10% resistor
1 8.2-V, 500-mW zener diode
1 DMM

- test leads and wires

PROCEDURE

Construct the circuit detailed in Fig. 5-14. The end of the zener with a dark band identifies the cathode end (lead) of the zener. This lead must connect to the 470-Ω resistor. Before connecting the power supply to the circuit, turn it on and adjust it for 12-V output. Then turn off the power supply; connect it to the circuit; turn on the power supply; and readjust for 12 V if necessary. Every time you make a change in the circuit, check the power supply output and readjust for 12 V if necessary.

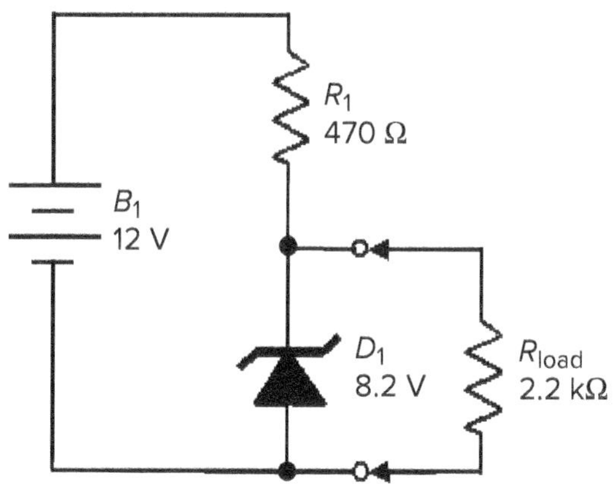

Fig. 5-14 Zener diode divider and regulator for experiment 5-8.

1. Measure and record the load voltage. ______________
2. **a.** How much voltage should be across R_1? ______________
   b. Measure and record V_{R_1}. ______________
   c. Was your answer to step a correct? ______________
3. Calculate and record the amount of current through the zener. ______________
4. Measure and record the current through the zener. ______________

5-10 LAB EXERCISE: SELF-DESIGNED CIRCUITS

PURPOSE

This exercise will give you experience in designing (and testing) circuits to meet specified criteria.

MATERIALS

Qty.		Qty.	
1	DMM and/or VOM	1	470-Ω, ½-W resistor
1	0- to 20-V dc power supply	2	1000-Ω, ½-W resistors
1	10-Ω ½-W resistor	1	2.2-kΩ, ½-W resistor
1	22-Ω, ½-W resistor	1	4.7-kΩ, ½-W resistor
1	51-Ω, ½-W resistor	2	10-kΩ, ½-W resistors
2	100-Ω, ½-W resistors	1	20-kΩ, ½-W resistor
1	220-Ω, ½-W resistor	•	test leads and wires

PROCEDURE

1. Using some combination of the resistors in the above materials list, design, construct, and test the circuits specified below. Remember not to exceed 50 percent of the power dissipation ratings of the resistors.
   a. A two-resistor circuit in which $V_{R_1} = 10V_{R_2}$ and $V_{R_1} = 14$ V.
   b. A two-resistor circuit in which $R_T = 68.75\ \Omega$ and $I_{R_1} = 2.2I_{R_2}$.
   c. A three-resistor circuit in which $I_{R_1} = I_{R_2}$ and $I_{R_1} = 2I_{R_3}$.
2. Write a formal report for this exercise that includes, as a minimum, schematic diagrams, measured and calculated values of electrical quantities, and possible reasons for disagreements between calculated and measured values.

NAME ____________________ DATE ____________

5-11 LAB EXPERIMENT: TROUBLESHOOTING—ONE FAULTY COMPONENT

PURPOSE

Electronic technicians are often required to troubleshoot multiple-load circuits and complex systems. This experiment will help you further develop your troubleshooting skills.

MATERIALS

Qty.

4 ½-W resistors (values known only to the instructor)
1 fuse, 0.5 A, with holder
1 20-V dc power supply
1 multimeter (VOM or DMM)
• test leads and wires
1 Multisim
1 Multisim file for Fig. 5-16

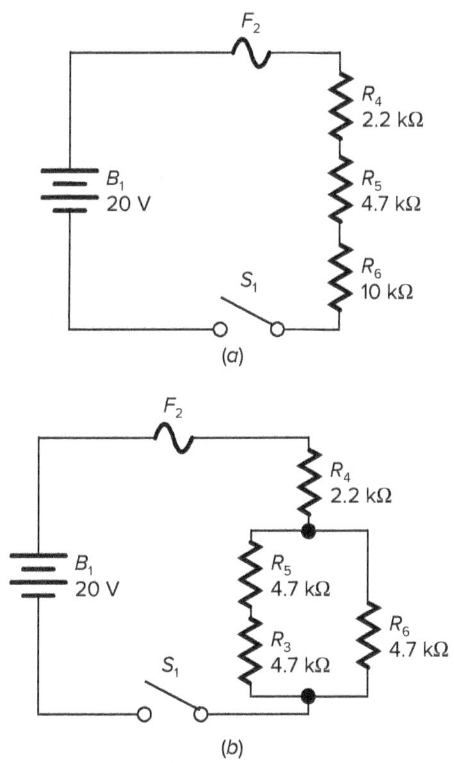

Fig. 5-16 Circuits for experiment 5-11.

PROCEDURE

1. The circuit for the diagram shown in Fig. 5-16(*a*) has one defective component. Unless your instructor asks you to use the Multisim file for Fig. 5-16, construct the circuit with the components provided by your instructor. Once the circuit is constructed, no components can be removed or disconnected to make measurements. (Pretend that they are soldered in place.) Use only the voltage and current functions of the multimeter.
   a. Which component is faulty? ____________

b. What is wrong with the component?

____

____

c. Submit a report explaining how you arrived at the answers to steps a and b. Include calculated values of I_T and R_T and the voltage drop for each resistor using the values shown on the diagram. Also include the measured value of I_T and each voltage drop.

2. Repeat step 1 for the circuit shown in Fig. 5-16(*b*).

a. Which component is faulty? ____________

b. What is wrong with the component?

____

____

c. Submit a detailed report as outlined in step 1c.

NAME ______________________________ DATE ______________

5-12 LAB EXPERIMENT: TROUBLESHOOTING—TWO FAULTY COMPONENTS

PURPOSE

This experiment will give you experience in locating two faulty components in the same circuit. It will illustrate that an understanding of fundamental circuit laws and relationships is essential when troubleshooting a circuit.

MATERIALS

Qty.

5 ½-W resistors (values known only to the instructor)
1 fuse, 0.5 A, with holder
1 20-V dc power supply
1 multimeter (VOM or DMM)
• test leads and wires
Multisim
Multisim file for Fig. 5-17

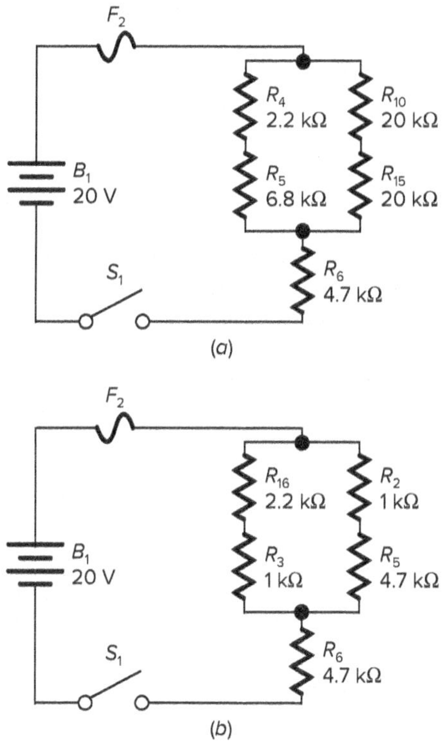

Fig. 5-17 Circuits for experiment 5-12.

PROCEDURE

1. The circuit for the diagram shown in Fig. 5-17(*a*) has two defective components. Unless your instructor asks you to use the Multisim file for Fig. 5-17, construct the circuit with the components provided by your instructor. Once the circuit is constructed, no components can be removed or disconnected to make measurements. (Pretend that they are soldered in place.) Use only the voltage and current functions of the multimeter.

a. Which components are faulty? ______________

b. What is wrong with the components?

____

____

c. Submit a report explaining how you arrived at the answers to steps a and b. Include calculated values of I_T and R_T and the voltage drop for each resistor using the values shown on the diagram. Include also the measured value of I_T and each voltage drop.

2. Repeat step 1 for the circuit shown in Fig. 5-17(*b*).

a. Which components are faulty? ______________

b. What is wrong with the components?

____

____

c. Submit a detailed report as outlined in step 1c.

NAME ____ DATE ____________________

CHAPTER 6

Complex–Circuit Analysis

TEST: COMPLEX-CIRCUIT ANALYSIS

For questions 1 to 5, supply the missing word or phrase in each statement.

1. The superposition theorem is applicable only to ____________ circuits.
2. Current is independent of load resistance for a(n) ____________ source.
3. ____________ theorem reduces a circuit to an ideal current source and a parallel resistance.
4. Application of ____________ theorem and ____________ theorem may not allow one to solve for all values of current and voltage.
5. ____________ independent equations are required to simultaneously solve three variables.

For questions 6 to 21, determine whether each statement is true or false.

6. When converting a Norton equivalent circuit to a Thevenin equivalent circuit, R_{TH} will be equivalent to R_N.
7. When a Norton equivalent circuit is converted to a Thevenin equivalent circuit, the V_{oc} of the two circuits will be the same.
8. The short-circuit current of a Thevenin equivalent circuit equals the constant current of a Norton equivalent circuit when both equivalent circuits are derived from the same complex circuit.
9. A three-loop complex circuit produces at least four unknown variables in the set of loop equations.
10. A complex circuit that requires two loops will have three unknown variables in at least one of the loop equations.
11. All loop equations for a given complex circuit will have the same number of unknown variables.
12. A Thevenin equivalent circuit represents an ideal (constant) voltage source.
13. When analyzing a three-source complex circuit by the superposition method, only one source is shorted out at any given time.
14. When the simultaneous solution of loop equations yields a negative value for a current, the assumed direction of the current is incorrect.
15. Application of the superposition theorem usually leads to the solution of all voltages and currents without the use of simultaneous equations.
16. Application of Norton's theorem always leads to the solution of all voltages and currents without the use of simultaneous equations.
17. When thevenizing a circuit, R_{TH} and V_{TH} can always be calculated with series-parallel circuit procedures.
18. All but one of the sources in a multiple-source complex circuit are shorted out when calculating I_N for the Norton equivalent circuit.

19. Some complex circuits can be analyzed using only series-parallel rules and procedures. 19. ______

20. If a single-source complex circuit is to be analyzed as a two-terminal network, the ends of the source are usually selected to be the terminals. 20. ______

21. If V_{RL} for many values of R_L must be determined, a circuit should be thevenized. 21. ______

Solve the following problems and record your answers. Be sure to specify units when appropriate.

22. What equation results from adding $-2x + 4y = 15$ to $4x + 2y + 4z = -8$? 22. ______

23. Solve for the values of A, B, and C when $2A - 5B = -4$, $-3B + 4C = 18$, and $3A + 4B - 2C = 5$. 23. ______

24. Calculate the value of R_{TH} and V_{TH} when the circuit in Fig. 6-1 is thevenized by using R_2 as the load resistor. 24. ______

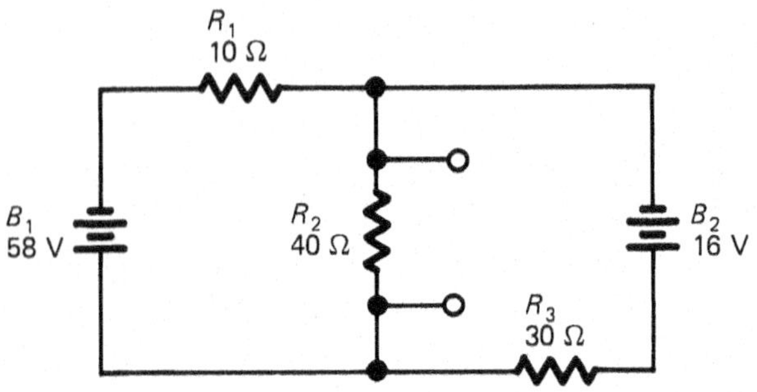

Fig. 6-1 Circuit for questions 24 and 25.

25. Use the superposition theorem to determine I_{R_3}, V_{R_3}, and I_{R_1} in the circuit shown in Fig. 6-1. 25. ______

26. Use the loop-equation technique to compute I_{R_2} and I_{R_3} for the circuit of Fig. 6-2. 26. ______

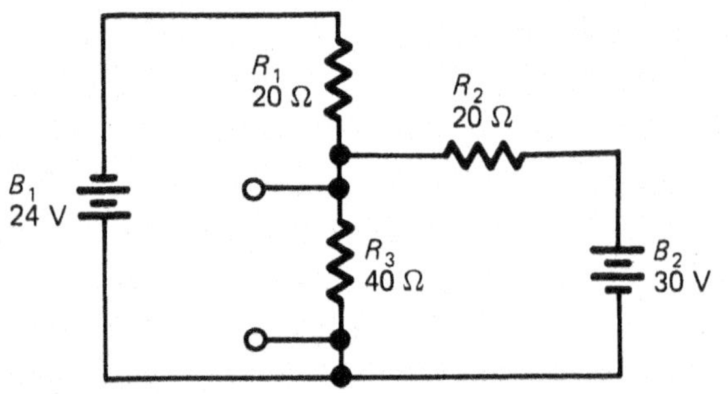

Fig. 6-2 Circuit for questions 26 to 28.

27. Calculate the value of I_N and R_N for the circuit shown in Fig. 6-2. (Do not assume that R_3 is the load resistor.) 27. ______

28. How much current will flow in a 50-Ω resistor connected between the terminals in Fig. 6-2? 28. ______

NAME ____ DATE ____________________

6-1 LAB EXPERIMENT: ANALYZING A BRIDGE CIRCUIT

PURPOSE

This experiment will provide experience in calculating and measuring currents and voltages in a complex circuit. Within the tolerances of the equipment and components, the measured values will verify whether or not the calculations are correct.

MATERIALS

Qty.		Qty.	
1	20-V dc power supply	1	4.7-kΩ, ½-W, 5% resistor
1	470-Ω, ½-W, 5% resistor	1	10-kΩ, ½-W, 5% resistor
1	1-kΩ, ½-W, 5% resistor	1	VOM or DMM
1	2.2-kΩ, ½-W, 5% resistor	•	test leads and wires

PROCEDURE

1. In Fig. 6-3, will point *A* be negative or positive with respect to point *B*? (Hint: Remove R_5 and determine whether point *A* or point *B* is more negative with respect to point *C*.) ______________

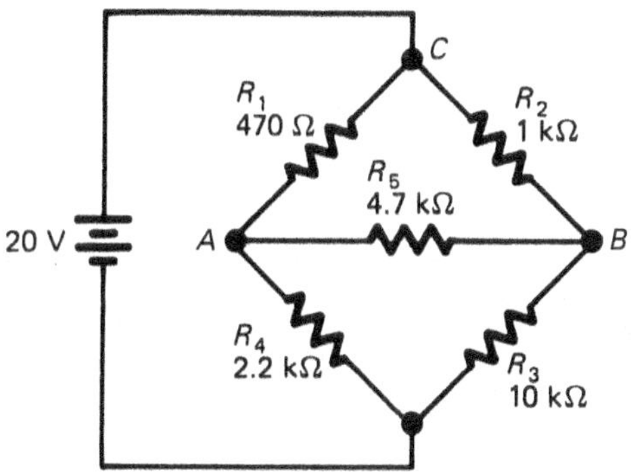

Fig. 6-3 Circuit for experiment 6-1.

2. Using Thevenin's theorem, calculate and record V_{R_5} and I_{R_5} for the circuit in Fig. 6-3. ______________ ______________
3. Construct the circuit shown in Fig. 6-3.
   a. Measure and record the voltage across R_5. ______________
   b. Is the polarity of V_{R_5} as you predicted in step 1? Is the value of V_{R_5} within ±10 percent of the value calculated in step 2? ______________ ______________
   c. Measure and record I_{R_5} and compare the measured value with the value calculated in step 2. If these two values differ by more than 10 percent, check your measurement technique and your calculations for error. ______________
   d. Knowing the value of I_{R_5} and V_{R_5}, can you determine other currents and voltages with just Kirchhoff's and Ohm's laws? ______________
   e. If you also knew V_{R_2}, in addition to V_{R_5} and I_{R_5}, could you then compute the other currents and voltages using only Kirchhoff's and Ohm's laws? ______________
4. Using Thevenin's theorem, compute and record the value of V_{R_2}, and I_{R_2} for Fig. 6-3. ______________ ______________

5. Verify your calculations in step 4 by measuring and recording V_{R_2} and I_{R_2}.

_______________ _______________

6. Compute and record I_{R_3} by applying Kirchhoff's current law to the junction at point B. Use the values of I_{R_5} and I_{R_2} from steps 2 and 4, respectively.

____

7. Measure and record I_{R_3}. Does the measured value match the value calculated in step 6? _______________ _______________
8. Compute and record V_{R_4} by applying Kirchhoff's voltage law to the loop that includes R_2, R_5, and R_4. _______________
9. Measure and record V_{R_4}. It should be within 10 percent of the value calculated in step 8. _______________
10. If you decreased the value of R_5 to 2.7 kΩ, would I_{R_5} increase or decrease?

____

11. Calculate and record the value of I_{R_5} when R_5 is changed to 6.8 kΩ. (Hint: Use the value of V_{TH} and R_{TH} determined in step 2.) _______________
12. Will changing the value of R_5 change the value of I_{R_4}? _______________

6-2 LAB EXPERIMENT: THE CONSTANT-CURRENT SOURCE

PURPOSE

This experiment will introduce you to a practical constant-current source that uses a transistor, a zener diode, and two resistors. It will illustrate the two major characteristics of a constant-current source: very high internal resistance and a load current that does not vary as the load resistance varies.

MATERIALS

Qty.		Qty.	
1	0- to 20-V dc power supply	1	4.7-kΩ, ½-W, 5% resistor
1	2N4126 transistor	1	10-kΩ, ½-W, 5% resistor
1	14-V, 1-W, 5% zener diode	1	20-kΩ, ½-W, 5% resistor
1	100-Ω, ½-W, 5% resistor	1	VOM or DMM
1	1-kΩ, ½-W, 5% resistor	•	test leads and wires
1	2.2-kΩ, ½-W, 5% resistor		

PROCEDURE

1. Construct the circuit shown in Fig. 6-4. Both Z_1 (the zener diode) and Q_1 (the transistor) are polarized devices, and so you must identify the leads before inserting them into the circuit. Refer to Fig. 6-5 to find out how to identify the leads.

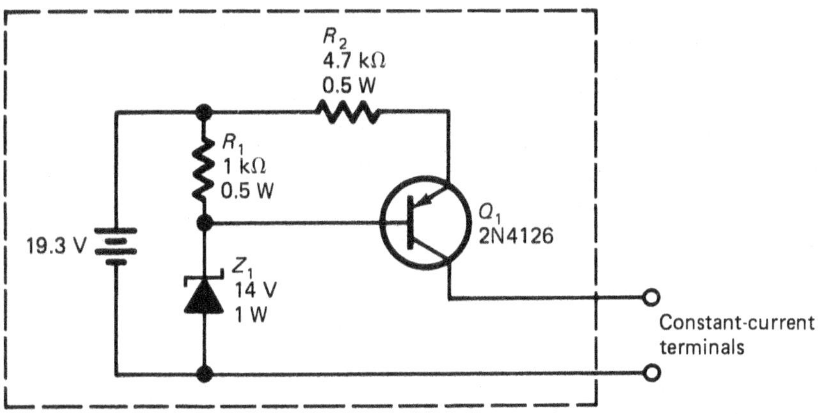

Fig. 6-4 Constant-current source for experiment 6-2. The dotted lines indicate that the complete circuit, including power supply, is the current source.

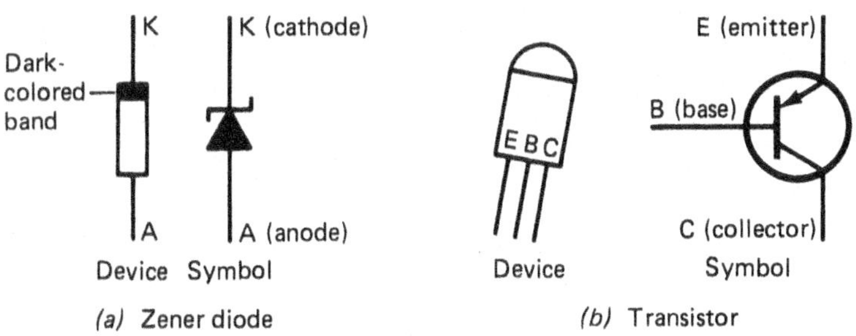

Fig. 6-5 Lead identification for the zener diode and transistor used in the constant-current source for experiment 6-2.

2. Temporarily disconnect the power supply from the circuit you just constructed. Then, using the DMM or the VOM on the ×1k range, measure and record the resistance between the constant-current terminals. Is the measured resistance approximately the value you would expect for a constant-current source? *Note:* The internal resistance of the source does change when power is applied, but for purposes of the experiment the change is not significant. ______________ ______________
3. Reconnect the power supply to the circuit and adjust it for 19.3 V. Now connect a 2.2-kΩ resistor to the terminals; then measure and record the current through the resistor. Also measure and record the voltage across the resistor. ______________ ______________
4. Replace the 2.2-kΩ resistor with a 100-Ω resistor, and again measure and record the current through and voltage across the resistor. Did the measured value of current change from step 3 to step 4? Did the measured value of voltage change from step 3 to step 4?

   ______________ ______________ ______________ ______________
5. If you now replace the 100-Ω resistor with a 10-kΩ resistor, how much current should the resistor draw? How much voltage would you expect across the 10-kΩ resistor? Check your answers by changing resistors and making the necessary measurements. ______________ ______________
6. Remove the 10-kΩ resistor and connect the milliammeter directly across the terminals of the source. Is the short-circuit current the same (within 5 percent) value as recorded in the previous three steps? ______________
7. You have probably noticed that constant-voltage sources, like cells and batteries, lose their constant-voltage characteristics when too much current is drawn from them. That is, the voltage at the terminals drops as the load resistance becomes too small.

   Constant-current sources also have limits. When the load resistance, and thus the load voltage, becomes too large, the source loses its constant-current characteristics. That is, the current from the terminals starts to decrease.

   Connect a 20-kΩ resistor to the current source; then measure and record the current through the 20-kΩ resistance. ______________
8. List four characteristics of a constant-current source that have been demonstrated by this activity.

   a. ______________________________

   b. ______________________________

   c. ______________________________

   d. ______________________________

NAME ______________________________ DATE ______________

6-3 LAB EXPERIMENT: EQUIVALENT CIRCUITS—NORTON'S AND THEVENIN'S

PURPOSE

This experiment will provide additional experience in converting complex circuits into equivalent circuits. It will also verify the relationships between Norton's equivalent circuit and Thevenin's equivalent circuit.

MATERIALS

Qty.		Qty.	
1	VOM or DMM	1	2N4126 transistor
1	6-V battery or dc power supply	1	2.2-kΩ, ½-W, 5% resistor
1	0- to 20-V dc power supply	2	4.7-kΩ, ½-W, 5% resistors
1	14-V, 1-W, 5% zener diode	1	10-kΩ, ½-W, 5% resistor
4	1-kΩ, ½- or 1-W, 2 or 5% resistors	1	20-kΩ, ½-W, 5% resistor
		•	test leads and wires

PROCEDURE

If your instructor asks you to perform this experiment using Multisim, record the measured values in the MS (Multisim) columns of Table 6-1. If you are using physical (real) components, record measured values in the HW (Hardware) columns.

1. Construct the complex circuit drawn in Fig. 6-6(*a*). Measure V_{R_L} and I_{R_L}. Record these values in Table 6-1 in the first row and the first and second columns. Now change R_L in Fig. 6-6(*a*) to 2.2 kΩ and again measure and record V_{R_L} and I_{R_L}. Finally, change R_L to 4.7 kΩ and measure and record V_{R_L} and I_{R_L}.

TABLE 6-1 Measured Currents and Voltages for the Circuits of Fig. 6-6

Type of Circuit	Value of R_L											
	1000 Ω				2200 Ω				4700 Ω			
	V_{R_L}		I_{R_L}		V_{R_L}		I_{R_L}		V_{R_L}		I_{R_L}	
	MS	HW	MS	HW	MS	HW	MS	HW	MS	HW	MS	HW
Complex [Fig. 6-6(*a*)]												
Norton's [Fig. 6-6(*b*) and (*c*)]												
Thevenin's [Fig. 6-6(*d*) and (*e*)]												

2. Using R_L as the load resistor, nortonize the circuit of Fig. 6-6(*a*). Record your calculated values of R_N and I_N.

   ______________ ______________

3. Measure and record I_N for Fig. 6-6(*a*) by removing R_L and replacing it with the DMM set on the 10-mA range. (The internal resistance of the DMM on the 10-mA range is very small compared with that of the other resistors in the circuit. Thus, it approximates a short circuit.) Does this measured I_N agree, within 10 percent, with the calculated I_N of step 2?

   ______________ ______________

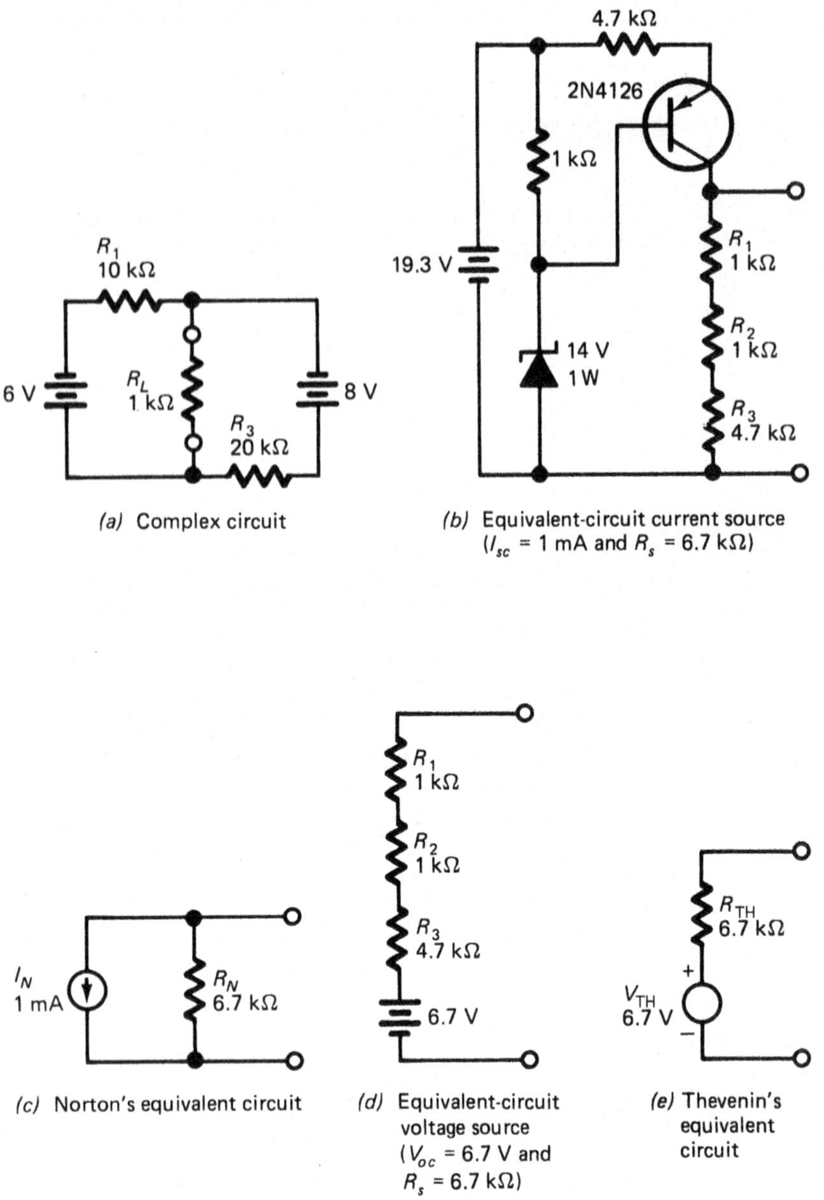

(a) Complex circuit

(b) Equivalent-circuit current source (I_{sc} = 1 mA and R_s = 6.7 kΩ)

(c) Norton's equivalent circuit

(d) Equivalent-circuit voltage source (V_{oc} = 6.7 V and R_s = 6.7 kΩ)

(e) Thevenin's equivalent circuit

Fig. 6-6 Circuits for experiment 6-3.

4. The Norton equivalent circuit you developed in step 2 should look like the one detailed in Fig. 6-6(*c*). If it doesn't, try again! Figure 6-6(*b*) shows a practical circuit diagram for the Norton equivalent circuit of Fig. 6-6(*c*). It is composed of the 1-mA constant-current source used in experiment 6-2 and an added 6.7 kΩ (R_1, R_2, and R_3) of internal parallel resistance. The circuits of Fig. 6-6(*b*) and (*c*) are electrically equivalent to that of Fig. 6-6(*a*) when R_L is removed from Fig. 6-6(*a*).
5. Should a 1-kΩ load across the terminals of Fig. 6-6(*b*) receive the same current and voltage as the 1-kΩ load (R_L) in Fig. 6-6(*a*)? ______________
6. Connect different values of R_L to Fig. 6-6(*b*) and make the measurements necessary to complete the second row of Table 6-1.
7. Does the data in rows 1 and 2 of Table 6-1 support Norton's theorem? ______________
8. Convert the Norton equivalent circuit you developed in step 2 into a Thevenin equivalent circuit. Record your calculated values of R_{TH} and V_{TH}.

______________ ______________

9. Reconstruct the circuit of Fig. 6-6(*a*). Remove R_L from the circuit. Then measure and record the voltage between the points from which you removed R_L. Is this voltage equal (within 10 percent) to V_{TH} calculated in step 8? ______________ ______________
10. Your Thevenin equivalent circuit from step 8 should look like the one in Fig. 6-6(*e*). A practical implementation of Fig. 6-6(*e*) is shown in Fig. 6-6(*d*). Construct the circuit of Fig. 6-6(*d*) and measure and record the voltage between its terminals. Should this voltage be equal to the voltage measured in step 9? ______________ ______________
11. Connect the necessary loads to the circuit of Fig. 6-6(*d*) and make the necessary measurements to complete the last row of Table 6-1. Do rows 2 and 3 of Table 6-1 indicate that Fig. 6-6(*b*) and (*d*) are equivalent circuits?

    ____
12. When R_L of Fig. 6-6(*a*) is removed, are all the circuits in Fig. 6-6 electrically equivalent in terms of providing current and voltage to a specific load?

    ____

6-4 LAB EXPERIMENT: THE SUPERPOSITION THEOREM

PURPOSE

In this experiment you will verify the techniques used in applying the superposition theorem. You will also gain additional experience in working with complex dual-source circuits.

MATERIALS

Qty.

1 6-V battery or dc power supply
1 0- to 20-V dc power supply
2 1-kΩ, ½-W, 5% resistors

Qty.

1 2.2-kΩ, ½-W, 5% resistor
1 VOM or DMM
• test leads and wires

PROCEDURE

1. Calculate and record the value and direction (left to right or right to left) of I_{R_1} in Fig. 6-7. ______________ ______________
2. Construct the circuit shown in Fig. 6-7 and then measure and record I_{R_1}. Did the measured value and direction agree with step 1?

______________ ______________

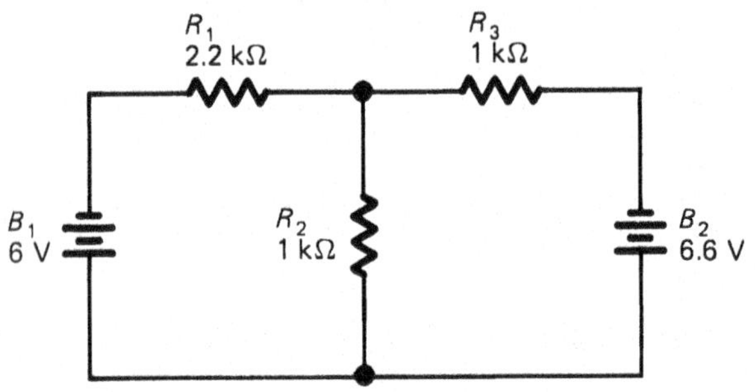

Fig. 6-7 Circuit for experiment 6-4.

3. Replace B_1 with a conductor and then measure I_{R_1}. Record both the direction and value of I_{R_1}. ______________ ______________
4. Remove the short (conductor) inserted in step 3, and place B_1 back in the circuit. Now replace B_2 with a short circuit, and once again measure and record the magnitude and direction of I_{R_1}.

______________ ______________

5. Is the direction of I_{R_1} in steps 3 and 4 the same? Is the magnitude of I_{R_1} in the two steps the same? ______________ ______________
6. Does the algebraic sum of I_{R_1} in step 3 and I_{R_1} in step 4 equal the value of I_{R_1} measured in step 2? ______________
7. Do your results support the superposition theorem? ______________
8. Will the direction of I_{R_2} be different when B_1 is shorted from when B_2 is shorted? ______________
9. If the polarity of B_2 in Fig. 6-7 is reversed, will the value of I_{R_1} increase or decrease?
10. Check your answer to step 9 by reversing B_2 and measuring and recording I_{R_1}. ______________

NAME ______________________ DATE ______________

6-5 LAB EXPERIMENT: ANALYSIS OF A VOLTAGE SOURCE

PURPOSE

Completion of this experiment will provide experience in experimentally developing a Thevenin equivalent circuit for an "unknown" voltage source.

MATERIALS

Qty.		Qty.	
1	"unknown" voltage source	1	470-Ω, ½-W resistor
1	VOM or DMM	1	1000-Ω, ½-W resistor
1	220-Ω, ½-W resistor		

DISCUSSION

The Thevenin equivalent circuit of a voltage source can be determined after making two voltage measurements. The first measurement determines the open-circuit terminal voltage (V_{oc}), and the second measurement determines the terminal voltage under loaded conditions (V_{R_L}). Figure 6-8 shows that the loaded voltage source can be treated as a series circuit. Thus, the voltage-divider equation can be applied to the circuit:

$$V_{R_L} = \frac{V_{oc} R_L}{R_L + R_S}$$

Solving this equation for R_S yields

$$R_S = \frac{V_{oc} R_L}{V_{R_L}} - R_L$$

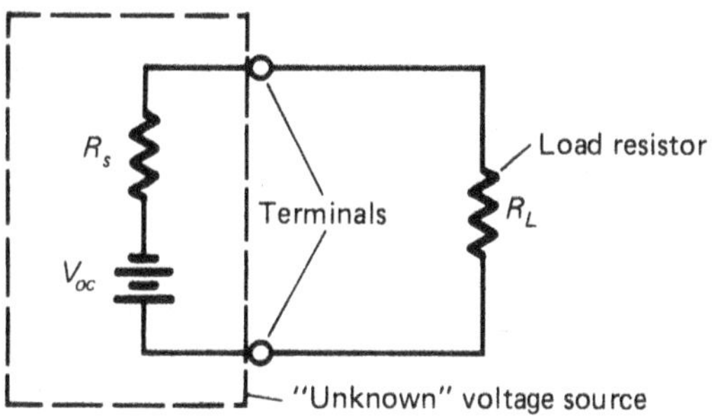

Fig. 6-8 Equivalent circuit of a loaded voltage source.

Since we know the value of R_L and can measure the values of V_{oc} and V_{R_L}, we can calculate the value of R_S using this equation. The values of R_S and V_{oc} are all that we require for the Thevenin equivalent circuit of the voltage source ($R_{TH} = R_S$ and $V_{oc} = V_{TH}$).

PROCEDURE

1. Measure and record the open-circuit voltage of the unknown voltage source.

   ____

2. Connect a 220-Ω, ½-W resistor to the voltage source. Measure and record the voltage across the resistor. Disconnect the resistor. ______________
3. Using the data obtained in steps 1 and 2, calculate and record the internal resistance of the voltage source. ______________

4. Draw a diagram of the Thevenin equivalent circuit of the voltage source.
5. Using the Thevenin equivalent circuit of step 4, predict—and record—the voltage across a 470-Ω resistor when the resistor is connected to the voltage source. ______________
6. Connect a 470-Ω resistor to the voltage source and measure—and record—the voltage across it. Does the measured voltage agree (±10 percent) with the predicted voltage of step 5? ______________ ______________
7. Convert the Thevenin equivalent circuit of step 4 to a Norton equivalent circuit. Record the value of the Norton current (I_N). ______________
8. Predict and record the current through a 1-kΩ resistor when it is connected to the unknown voltage source. ______________
9. Using the 10-mA range of the DMM, measure and record the current through a 1-kΩ resistor when it is connected to the voltage source. Does the measured current agree (±10 percent) with the current predicted in step 8? ______________ ______________

NAME ______________________________ DATE ______________

6-6 LAB EXPERIMENT: TROUBLESHOOTING A BRIDGE

PURPOSE

This experiment provides troubleshooting experience on a complex circuit. It can be completed using Multisim with a fault inserted by your instructor.

MATERIALS

Multisim
Multisim file for Fig. 6-9

PROCEDURE

1. The Wheatstone bridge shown in Fig. 6-9 has one defective resistor. Your problem is to determine which resistor it is without disconnecting any components. (Assume that all parts are soldered in place.)
2. Which resistor is defective? ______________
3. Submit a report detailing how you determined which resistor was faulty.

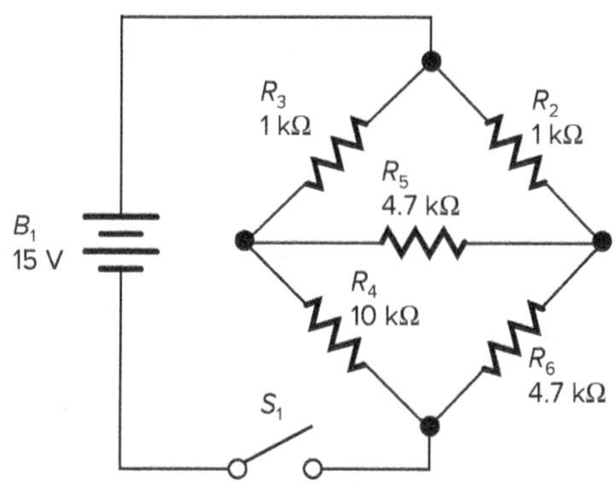

Fig. 6-9 Circuit for experiment 6-6.

6-7 LAB EXPERIMENT: ANALYZING AND TROUBLESHOOTING COMPLEX CIRCUITS

PURPOSE

One must know what to expect of a circuit before determining if it is faulty or trying to troubleshoot it. This experiment provides experience in determining if a circuit is faulty and troubleshooting the circuit if it is faulty.

MATERIALS

Qty.

Multisim
Multisim files for Fig.s 6-10 and 6-11.

PROCEDURE

1. In the circuit of Fig. 6-10 have your instructor insert a fault.
   a. Analyze the circuit and record the calculated values for

      I_{R_1} = ______________

      I_{R_2} = ______________

      I_{R_3} = ______________

      V_{R_2} = ______________

      V_{R_3} = ______________

   b. Determine and list the faulty component. ______________
   c. In what way is the component faulty?______________

      ____

   d. Submit a report explaining how you arrived at your answers to steps b and c above.

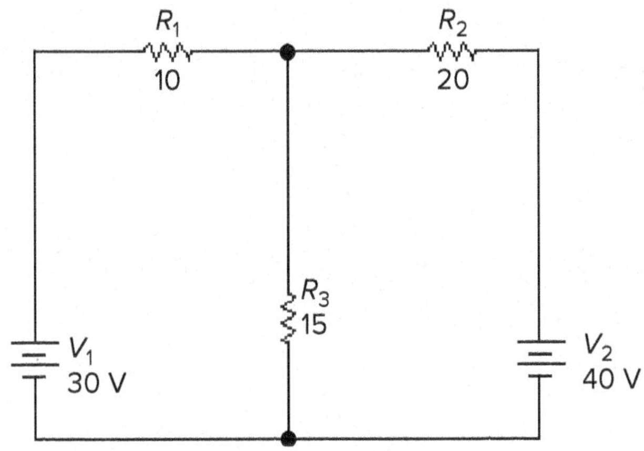

Fig. 6-10 Circuit for experiment 6-7.

2. Repeat step 1 for the circuit in Fig. 6-11.
   a. I_{R_1} = ______________

      I_{R_2} = ______________

      I_{R_3} = ______________

      V_{R_2} = ______________

      V_{R_3} = ______________

   b. ______________

   c. ______________

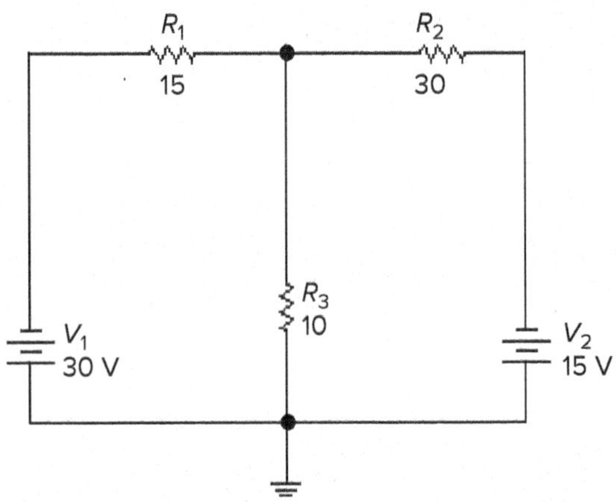

Fig. 6-11 Circuit for experiment 6-7.

NAME ____ DATE ________________

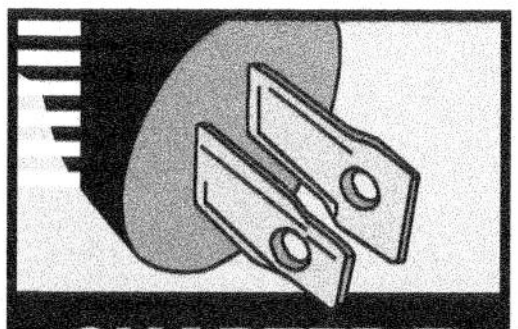

CHAPTER 7

Magnetism and Electromagnetism

TEST: MAGNETISM AND ELECTROMAGNETISM

For questions 1 to 24, determine whether each statement is true or false.

1. Invisible magnetic lines of force surround all materials. **1.** ____________
2. All the flux from the north pole of a bar magnet extends out parallel to the bar magnet. **2.** ____________
3. The flux from a horseshoe magnet is least dense halfway between its poles. **3.** ____________
4. Flux leaves the south pole and enters the north pole of the magnet. **4.** ____________
5. Magnetic poles are attracted to each other when their fluxes join together. **5.** ____________
6. A magnet's field is usually destroyed when the magnet is broken into two pieces. **6.** ____________
7. Continuous loops of flux pass through a bar magnet as well as through the surrounding air. **7.** ____________
8. A circular magnet can have both a north and a south pole. **8.** ____________
9. Any current-carrying conductor has a magnetic field around it. **9.** ____________
10. The current in a coil is flowing clockwise when viewed from the bottom of the coil. The top of the coil is a north pole. **10.** ____________
11. Nonmagnetic materials block magnetic flux. **11.** ____________
12. A magnetic field is associated with a neutron. **12.** ____________
13. Magnetic domains are present in all materials. **13.** ____________
14. The domains in a magnetized material are aligned in an orderly fashion. **14.** ____________
15. Tin is a temporary magnetic material. **15.** ____________
16. The insertion of any magnetic material in the center of a coil carrying direct current increases the pole flux. **16.** ____________
17. Permanent magnetic materials can be either magnetized or demagnetized with a dc coil. **17.** ____________
18. A piece of steel has more reluctance than a piece of brass of equal size. **18.** ____________
19. Current changes in one coil can induce a voltage in a second coil. **19.** ____________
20. An electromagnet's strength is increased when its core is replaced with a higher-permeability material. **20.** ____________
21. The magnetic polarity of the armature in a dc motor changes as the motor rotates. **21.** ____________
22. Relays can be made to operate on either alternating or direct current. **22.** ____________
23. The movable part of a solenoid is made from nonmagnetic material. **23.** ____________
24. Electromechanical devices used to operate mechanical devices like brakes and clutches are called solenoids. **24.** ____________

For questions 25 to 36, supply the missing word or phrase needed in each statement.

25. A(n) _______________ is a natural permanent magnet. 25. _______________
26. The lines of force in a magnetic field are called _______________. 26. _______________
27. The north pole of a magnet attracts the _______________ pole of another magnet. 27. _______________
28. Parallel conductors with currents flowing in the same direction _______________ each other. 28. _______________
29. When other factors are unchanged, increasing the space between the turns of a coil causes the pole flux to _______________. 29. _______________
30. A conductor has no induced voltage if it moves _______________ to magnetic flux. 30. _______________
31. The permeability of a material _______________ as the material saturates. 31. _______________
32. _______________ magnetic material is used for the core of electromagnets. 32. _______________
33. Ampere-turn is the base unit of _______________. 33. _______________
34. Weber is the base unit for _______________. 34. _______________
35. Tesla is the base unit for _______________. 35. _______________
36. The part of a dc motor that makes contact with the brushes is the _______________. 36. _______________

For questions 37 to 47, choose the letter that best completes each statement.

37. Halving the distance between magnetic poles makes the force between the poles 37. _______________
    - **a.** Twice as great
    - **b.** Four times as great
    - **c.** Half as great
    - **d.** One-fourth as great
38. Current flowing in a straight conductor produces a magnetic field with 38. _______________
    - **a.** A north pole at the negative end of the conductor
    - **b.** A north pole at the positive end of the conductor
    - **c.** No north pole or south pole
    - **d.** A flux at 45° to the conductor
39. You are viewing the end of a conductor in which the current is flowing away from you. The direction of the magnetic flux of the conductor is 39. _______________
    - **a.** Counterclockwise
    - **b.** Clockwise
    - **c.** Away from you
    - **d.** Toward you
40. Which of the following elements is magnetic? 40. _______________
    - **a.** Lead
    - **b.** Tin
    - **c.** Copper
    - **d.** Iron
41. Magnetic shields are made from 41. _______________
    - **a.** Nonmagnetic materials
    - **b.** High-reluctance materials
    - **c.** Low-reluctance materials
    - **d.** Nonmetallic materials
42. Field intensity can be increased by 42. _______________
    - **a.** Increasing the current in a coil
    - **b.** Decreasing the reluctance of the magnetic circuit
    - **c.** Decreasing the number of turns in a coil
    - **d.** Increasing the permeability of the magnetic circuit
43. When a magnetic material is saturated, 43. _______________
    - **a.** Its reluctance is very low
    - **b.** Its permeability is very high
    - **c.** It is passing as much flux as it can
    - **d.** Its flux density approaches zero

44. The opposition to magnetic flux is called **44.** ______
- **a.** Flux density
- **b.** Residual magnetism
- **c.** Reluctance
- **d.** Permeability

45. The movable part of a solenoid is called the **45.** ______
- **a.** Armature
- **b.** Coil
- **c.** Commutator
- **d.** Plunger

46. The movable part of a relay is called the **46.** ______
- **a.** Armature
- **b.** Commutator
- **c.** Field
- **d.** Plunger

47. The rotating part of a dc motor is called the **47.** ______
- **a.** Armature
- **b.** Coil
- **c.** Field
- **d.** Brush

Answer the following questions.

48. Determine the flux density at the pole of a bar magnet which is 0.1 m long, 0.03 m wide, and 0.01 m high and produces 0.12 base unit of flux. **48.** ______

49. What is the mmf of a 500-turn coil when the coil draws 1.6 A? **49.** ______

50. Determine the magnetic field strength of a circuit which has 0.6 A in a 200-T coil on a core with an average length of 0.4 m. **50.** ______

51. What is the permeability of a material if the magnetizing force is 120 base units and the flux density is 0.4 base unit? **51.** ______

52. The mmf of a 450-turn coil is 585 base units. Determine the coil current. **52.** ______

53. Air has a permeability of 1.26×10^{-6} Wb/(A . t . m). What is the relative permeability (μ_r) of a magnetic material that has a permeability (μ) of 5.67×10^{-4} Wb/(A . t . m)? **53.** ______

2. Prepare the circuit shown in Fig. 7-2(*a*), but do not connect the lead to the negative terminal of the battery. One end of the wire arches can be connected by twisting the 36-gage wires together. Notice in Fig. 7-2(*a*) that the arch wires are electrically in series but physically parallel to each other.
   a. Will the current in the arch wires be in the same or opposite directions?

   ____

   b. Will the two wires attract or repel each other when the circuit is completed? ______________
   c. Check your answer to step 2b by touching the connecting lead to the negative terminal of the battery. Do the wires repel or attract each other?

   ____

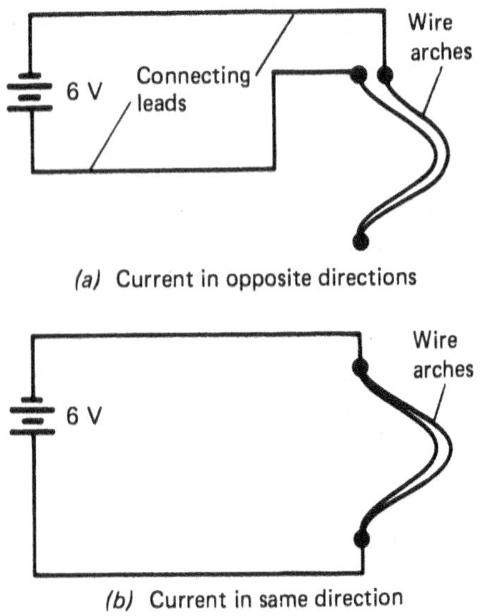

Fig. 7-2 Circuits for experiment 7-1.

3. Prepare the circuit of Fig. 7-2(*b*), but leave one lead off the battery terminal. The arch wires can be connected in parallel by twisting together the wires at each end.
   a. Are the currents in the arch wires in the same direction?

   ____

   b. Will the wires repel each other when the circuit is completed?

   ____

   c. Check your response to step 3b by momentarily connecting the circuit. Do the wires attract or repel each other? ______________
   d. What causes the arch wires to repel or attract each other?

   ____

   ____

4. A two-conductor cable supplies current to a load such as a large lamp. Will the conductors repel or attract each other? ______________ Why?

   ____

   ____

NAME ____ DATE ______________

7-2 LAB EXPERIMENT: MAGNETIC MATERIALS, POLES, AND FIELDS

PURPOSE

This experiment illustrates the existence of magnetic poles and fields. You will identify magnetic and nonmagnetic materials and determine flux concentration in high-permeability materials. As an option, this experiment may be more suitable as a classroom or laboratory demonstration led by an instructor.

MATERIALS

Qty.		Qty.	
2	magnets, bar, approximately 0.5 × 1 × 3 cm (3/16 × 3/8 × 1¼ in.)	1	sheet of paper, 8½ × 11 in.
1	salt shaker filled with iron filings	3	masking tape strips, 6 cm (2½ in.) long
1	ferrite rod, approximately 1 cm (3/8 in.) in diameter and 1.5 cm (5/8 in.) long	1	screwdriver
		•	small samples of iron (nails), copper, brass, aluminum, paper, wood, etc.

PROCEDURE

1. Using one of the bar magnets, determine which materials are magnetic and which are nonmagnetic. Also test the ferrite rod.

  a. List the magnetic materials.

  ____

  ____

  b. Will the ferrite rod attract iron (nails)? ______________

  c. Is ferrite a permanent or temporary magnetic material? ______________

2. Stroke the blade of the screwdriver with one end of a bar magnet.

  a. Will the screwdriver now attract magnetic materials? ______________

  b. What type of magnetic material is the screwdriver made of? ______________

3. Lay one of the magnets on the table and cover it with a sheet of paper. Sprinkle a thin coating of iron filings on the paper and gently tap the paper.

  a. Are the iron filings concentrated at the ends of the poles? ______________

  b. In the margin of this page, sketch the pattern made by the iron filings. Remove the lid from the salt shaker and pour the iron filings back into the shaker.

4. Tape one of the bar magnets to the top of a table or desk. Bring one end of the other magnet up to one end of the taped-down magnet. Note the force of either repulsion or attraction. Reverse the ends of the free magnet and note the opposite force.

  a. When the magnets repelled each other, were the poles of the same or opposite polarities? ______________

  b. Arrange and tape the two magnets, as shown in Fig. 7-3. The magnets should be about 1 cm (3/8 in.) apart, with unlike poles facing each other. Now cover the magnets with paper and sprinkle on iron filings. Tap the paper gently. Does the arrangement of the filings indicate that lines of force from the two poles join together? ______________ Sketch the pattern of the filings around the two poles. Pour the filings back into the shaker.

c. Reverse either of the two magnets so that like poles are facing each other. Again tape the magnet so that the poles are about 1 cm apart. Put the paper back over the magnets and sprinkle on a thin coat of filings. Tap lightly. Do the filings indicate that the flux from like poles joins together? ______________ In the margin of this page, sketch the pattern of the filings. Return the filings to the shaker.

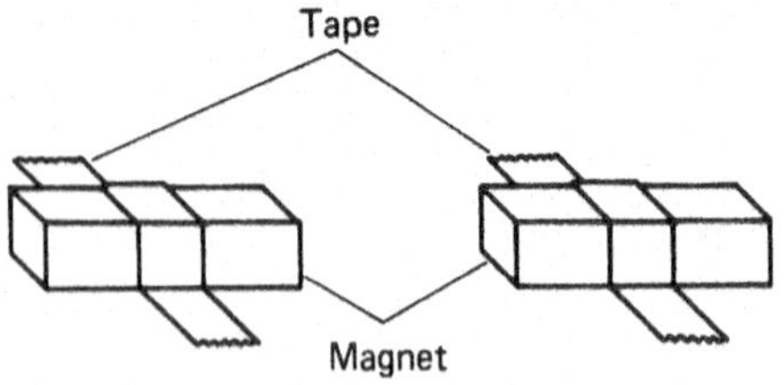

Fig. 7-3 Position of taped-down magnets for experiment 7-2.

5. Arrange the magnets and the ferrite rod, as shown in Fig. 7-4. Allow about 2 mm between the ferrite and the magnets. Have unlike poles of the magnet facing the ferrite rod. Cover the magnets and rod with paper. Sprinkle with iron filings and tap lightly.

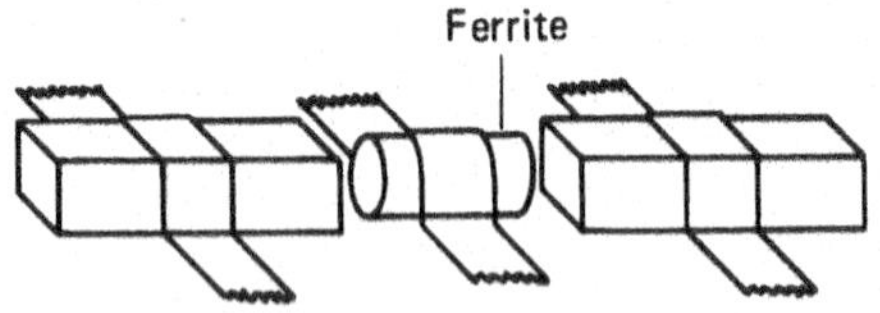

Fig. 7-4 Location of magnets and ferrite rod for experiment 7-2.

a. In the margin of this page, sketch the arrangement of the filings around the ferrite and the poles. Put the filings back in the shaker. Does the filing pattern indicate that the ferrite rod carried most of the flux from one pole to the other? ______________

b. Does ferrite have a smaller or larger permeability than air? ______________

c. Which has more reluctance, the ferrite rod or the air it replaces? ______________

NAME ______________________________ DATE ______________

7-3 LAB EXPERIMENT: ELECTROMAGNETISM

PURPOSE

The objectives of this experiment are to verify some principles of electromagnetism and to magnetize and demagnetize materials. Completing the experiment will provide experience in working with magnet wire. As an option, this experiment may be more suitable as a classroom or laboratory demonstration led by an instructor.

MATERIALS

Qty.		Qty.	
1	wooden dowel, ¼ in. in diameter × 4 in. long	1	6- to 8-V, 2-A dc power supply
1	150-ft magnet wire, 23 gage	1	sheet of paper, 8½ × 11 in.
1	screwdriver, approximately 3⁄16 × 4-in. blade	12	nails, 3d (3 penny), common
2	no. 47 lamps with holders	3	nails, 16d
1	6- to 8-V, 2-A ac power supply (a 6-V transformer will do)	1	resistor, 2 Ω, 10 W
		1	soldering gun
		•	masking tape, connecting wires

PROCEDURE

CAUTION **The soldering gun tip will heat up as you continue this experiment. Do not touch the tip when magnetizing or demagnetizing materials.**

1. Put the blade of the screwdriver between the heavy copper leads that form the tip of the soldering gun so that the tip leads surround the blade. Turn the soldering gun on and then off immediately. Remove the screwdriver and see if it will attract the small nails. If it does not, repeat the process.
   a. Has the screwdriver been magnetized? ______________
   b. Is it a permanent magnet? ______________
   c. Again insert the screwdriver while the gun is off. Turn the gun on and remove the screwdriver while the gun is on. Then turn off the gun. Should the screwdriver now be magnetized or demagnetized? ______________
   d. Try to pick up the 3d nails with the screwdriver. Does the screwdriver attract any nails? ______________
2. Demagnetize one of the large (16d) nails. Wind a coil on the nail. The coil should be about two-thirds the length of the nail. It should consist of four layers of 23-gage magnet wire. Leave about 25-cm (10-in.) leads on the start and finish ends of the coil. Use masking tape to hold each layer of wire in place. The construction details of the coil are shown in Fig. 7-5. When you have finished winding the electromagnet, remove about 1 cm (⅜ in.) of insulation from the ends of the coil. Now connect the circuit shown in Fig. 7-6. In this figure the two straight lines by the coil represent the iron core (nail) on which the coil is wound. The no. 47 lamp limits the current in the circuit to about 100 mA.
   a. Will the electromagnet you have just made attract small nails? ______________ Why?
   ____
   ____

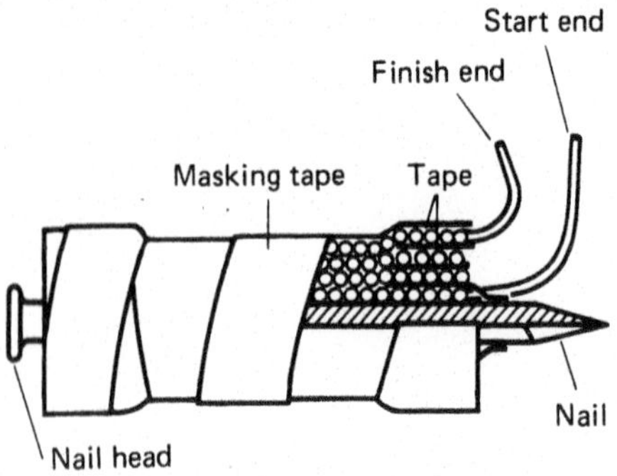

Fig. 7-5 Making an electromagnet.

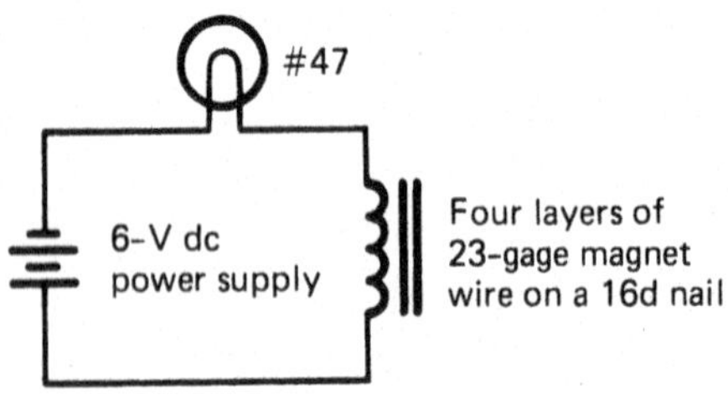

Fig. 7-6 Circuit for experiment 7-3, step 2.

b. Pick up as many small nails as you can with the electromagnet. While holding the electromagnet (and nails) above the table top, disconnect the electric circuit. What happened? ______________ Why?

____

____

c. Try picking up the small nails again while the electric circuit is disconnected. Is there any magnetism left in the electromagnet? ______________

d. What is the name for the magnetism left in a temporary magnet after the magnetizing force has been removed? ______________

CAUTION **Operate the following circuit only for short periods of time. The current in the coil will be about 2 A, which will cause the coil to heat up.**

3. Increasing the current in a coil will increase the magnetomotive force (and thus the strength) of an electromagnet.

a. Connect your electromagnet as shown in Fig. 7-7(*a*). See how many small nails your electromagnet will pick up now. Has the strength of your electromagnet increased? ______________

b. Disconnect the circuit and check the residual magnetism. Is it any greater than it was before? ______________ Why?

____

____

c. Now change from a dc to an ac power supply, as shown in Fig. 7-7(*b*). Connect the circuit, and see how many nails the magnet will pick up when powered by alternating current. Disconnect the circuit and check the residual magnetism. Is there a noticeable difference in the strength of the magnet when powered by alternating or direct current? ______________

d. Is there any difference in the residual magnetism? ______________

NAME ______________________________ DATE ______________

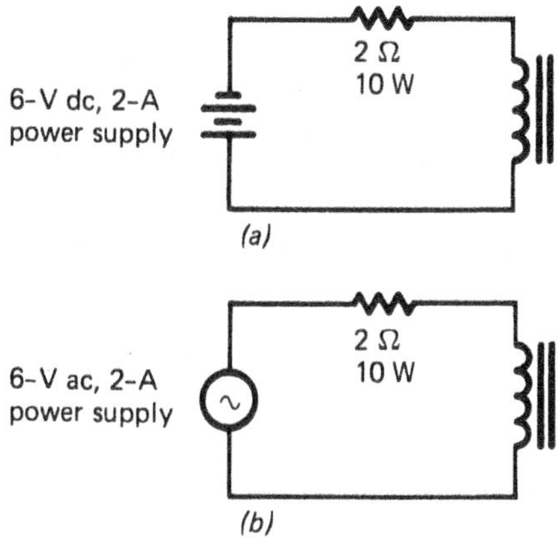

Fig. 7-7 Circuit for experiment 7-3, step 3.

4. Either construct or borrow a second electromagnet. Connect the two magnets in series and power them, as shown in Fig. 7-8. Using the left-hand rule, predict whether the heads of the two magnets will attract or repel.

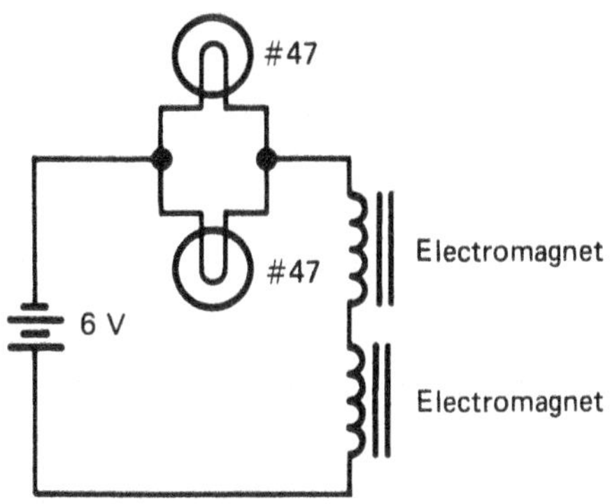

Fig. 7-8 Circuit for experiment 7-3, step 4.

a. Bring the heads of the magnets together. Do they attract or repel? ______________

b. Now reverse the direction of current in one of the magnets by interchanging the start and finish ends of one magnet. Again bring the heads of the magnets together. Do they now attract or repel each other? ______________

c. If you reverse the direction of current in the other magnet, would the heads repel or attract each other? ______________ Check your answer by reversing current in the second coil.

5. Now you are going to make a coil in which you can use different core materials. The coil will be wound on a paper form. Both the form and the coil will be wound on a ¼-in. dowel. Refer to Fig. 7-9 and fold a sheet of paper into four layers as shown. Next roll and tape the paper around the dowel, as shown in Fig. 7-10. Do not tape the paper to the dowel because the dowel will be removed later. Wind four layers of 23-gage magnet wire on top of the paper. Tape, as shown in Fig. 7-11, both ends of each layer with masking tape. When the coil is finished, remove the insulation from the ends of the magnet wire.

a. Connect the circuit of Fig. 7-12(*a*) and try to pick up the small nails with the wood-core magnet. Does the magnet attract the nails? ______________

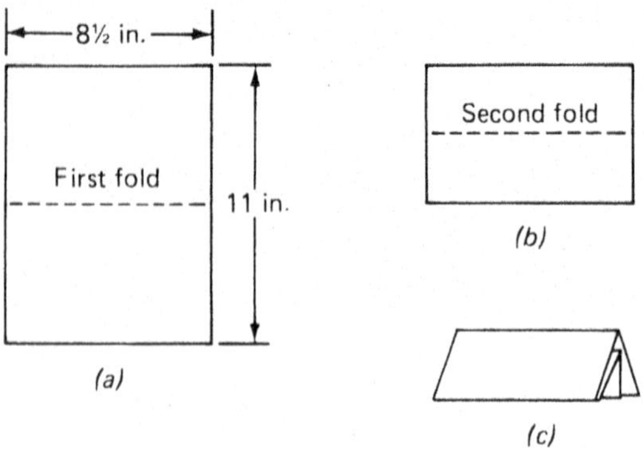

Fig. 7-9 Folding paper for coil form used in experiment 7-3, step 5.

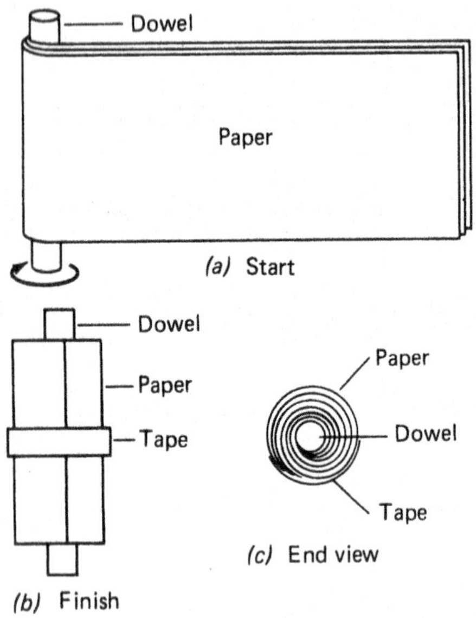

Fig. 7-10 The coil form is made by rolling the folded paper around the dowel.

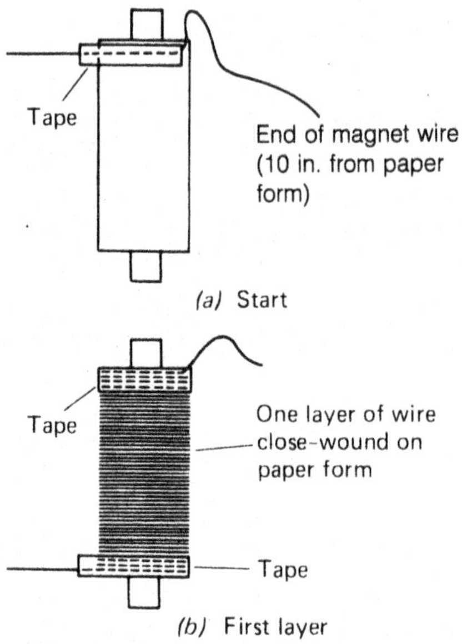

Fig. 7-11 Winding coil for experiment 7-3, step 5.

b. Remove the ¼-in. dowel by gently pulling and twisting on it. Now try to attract the small nails with the air-core magnet. Does it attract any better than the wood core? _______________ Why?

____

____

c. Insert a large (16d) nail into the core of the coil. Now try attracting the small nails with the electromagnet. Does it attract any better than either a wood-core or an air-core magnet? _______________ Why?

____

____

d. Remove the large nail from the core. Change the electromagnet circuit from that shown in Fig. 7-12(*a*) to that shown in Fig. 7-12(*b*). The current in the coil in Fig. 7-12(*b*) is about 10 times as great in Fig. 7-12(*a*). Now try to attract the small (3d) nails. Disconnect the circuit so that the coil does not overheat. Will the magnet now attract the small nails? _______________ Why?

____

____

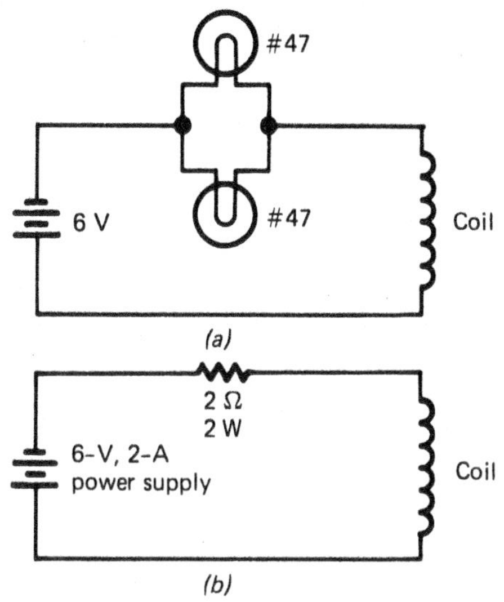

Fig. 7-12 Circuits for experiment 7-3, step 5.

e. Insert a large nail about one-third of the way into the coil. Now reconnect the circuit. What happens?

____

____

f. Why did this happen?

____

____

g. Hold the coil, with the nail in it and its head facing down, above the desk. Now disconnect the circuit. What happened?

____

____

h. Why? (What happened to the mmf?)

____

____

i. Could this coil be used to magnetize and demagnetize a screwdriver? ____________ Try it!

j. How could the circuit in Fig 7-12 be changed so that it could demagnetize a screwdriver?

____

____

NAME ______________________________ DATE ______________

CHAPTER 8

Alternating Current and Voltage

TEST: ALTERNATING CURRENT AND VOLTAGE

For questions 1 to 21, determine whether each statement is true or false.

1. Alternating current reverses the direction in which it is flowing at the end of each alternation. **1.** ________

2. Fluctuating direct current periodically touches the zero reference line of a waveform graph. **2.** ________

3. One hertz is equal to two alternations per second. **3.** ________

4. Electric power is distributed at 120 Hz in the United States. **4.** ________

5. Effective voltage and rms voltage are interchangeable terms. **5.** ________

6. The mechanical and electrical degrees of a two-pole generator are equal. **6.** ________

7. The output voltage of a generator can be changed by changing the flux density of its field. **7.** ________

8. The frequency of a generator can be changed by changing the number of poles in the field. **8.** ________

9. Both the frequency and the voltage of a generator change when the armature speed changes. **9.** ________

10. The rms voltage from a 120-V outlet is greater than the average voltage. **10.** ________

11. An ac generator uses a commutator and brushes. **11.** ________

12. Alternating current is used to excite the electromagnetic field coils of an ac generator. **12.** ________

13. A three-phase system requires only three conductors for its power line. **13.** ________

14. Each phase in a three-phase system is separated 120° from the other two phases. **14.** ________

15. The three phases of a three-phase system never have the same instantaneous polarity. **15.** ________

16. The line current and phase current are equal in a four-wire wye system. **16.** ________

17. The common connection in a wye system is called the star point. **17.** ________

18. The reference ends of the phase windings are connected together in a delta-connected system. **18.** ________

19. The neutral wire of a three-phase, four-wire system carries current under all load conditions. **19.** ________

20. Three-phase systems are more efficient than single-phase systems. **20.** ________

21. Rectified three-phase alternating current is closer to pure direct current than rectified single-phase alternating current. **21.** ________

For questions 22 to 31, supply the missing word or phrase in each statement.

22. The symbol or abbreviation for period is ________. **22.** ________________

23. The symbol or abbreviation for frequency is ________. **23.** ________________

24. The base unit of frequency is ________. **24.** ________________

25. The abbreviation or symbol for the base unit of frequency is ______________. 25. ______________

26. ______________ is the rate at which cycles are produced. 26. ______________

27. When a voltage waveform is plotted, voltage is usually plotted on the ______________ axis. 27. ______________

28. Either time or ______________ can be plotted on the ______________ axis of a waveform graph. 28. ______________ ______________

29. The time required to complete one cycle is the ______________. 29. ______________

30. There are ______________ degrees per cycle. 30. ______________

31. The line voltage is ______________ times higher than the phase voltage in a(n) ______________ system. 31. ______________ ______________

For questions 32 to 35, choose the letter that best completes each statement.

32. Commercial power is distributed as a 32. ______________
 a. Circle wave
 b. Triangle wave
 c. Sine wave
 d. Square wave

33. Which of the following does not influence the amount of voltage induced in a conductor? 33. ______________
 a. Flux density
 b. Direction the conductor moves
 c. Speed the conductor moves
 d. Angle at which the conductor cuts the flux

34. A given value of direct current has the same heating effect as an equal value of 34. ______________
 a. Peak ac current
 b. Peak-to-peak ac current
 c. Root-mean-square ac current
 d. Average ac current

35. A three-phase system that provides both 120-V single-phase and 208-V three-phase is the 35. ______________
 a. Four-wire delta system
 b. Four-wire wye system
 c. Five-wire delta system
 d. Five-wire wye system

For questions 36 to 44, solve each problem.

36. What is the period of a 500-Hz sine wave? 36. ______________

37. What is the frequency of a waveform that has a period of 0.4 ms? 37. ______________

38. What is the peak-to-peak value of 100 V? 38. ______________

39. What is the rms value of 180 V_{av}? 39. ______________

40. The rms current in a circuit is a 12 A. What is the peak current? 40. ______________

41. What is the frequency of a six-pole generator that rotates at 1000 r/min? 41. ______________

42. What is the line voltage of a wye-connected, three-phase generator that has a phase voltage of 260 V? 42. ______________

43. What is the phase current of a 120-V delta-connected, three-phase generator connected to three delta-connected, 40-Ω resistors? 43. ______________

44. What is the line current for the circuit described in question 43? 44. ______________

8-1 LAB EXERCISE: AC WAVEFORMS AND TERMINOLOGY

PURPOSE

This experiment will provide you with more experience in using waveforms to describe ac currents and voltages.

PROCEDURE

1. Figure 8-1(*a*) lists the voltages of an ac waveform for each 15° of the cycle. Using these voltages and corresponding degrees, plot the points of a waveform on the graph in Fig. 8-1(*b*).

Electrical degrees	Volts	Electrical degrees	Volts
0	0.0	195	-18.1
15	18.1	210	-35.0
30	35.0	225	-49.5
45	49.5	240	-60.6
60	60.6	255	-67.6
75	67.6	270	-70.0
90	70.0	285	-67.6
105	67.6	300	-60.6
120	60.6	315	-49.5
135	49.5	330	-35.0
150	35.0	345	-18.1
165	18.1	360	0.0
180	0.0		

(a) Data

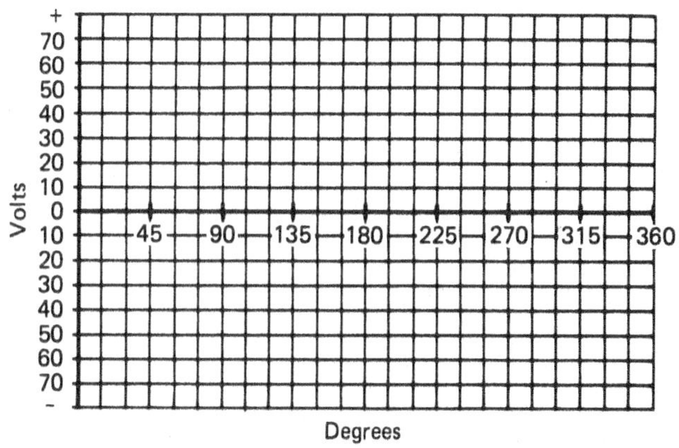

(b) Graph

Fig. 8-1 Plotting a sine wave.

a. Now connect the points together to finish the waveform.
b. What type of waveform does it make? ______________
c. Assume each degree of the cycle represents 0.0001 s. What is the period (*T*) of the waveform? ______________
d. What is the frequency of the waveform? ______________
e. What is the rms value of the waveform? ______________
f. What is the average value of the waveform? ______________

2. Refer to Fig. 8-2.
a. What type of waveform is shown? ______________
b. What is the peak-to-peak value of the waveform? ______________
c. What is the peak positive value of the waveform? ______________
d. What is the period of this waveform? ______________
e. What is its frequency? ______________
f. How many cycles of this waveform will be completed in 1 s? ______________

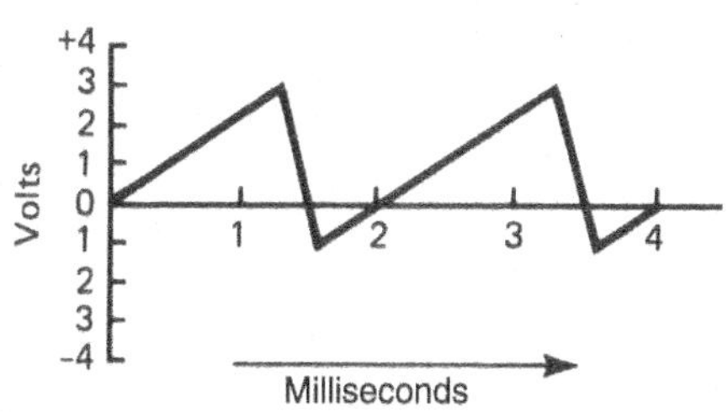

Fig. 8-2 Waveform for experiment 8-1, step 2.

8-2 LAB EXERCISE: THREE-PHASE CONNECTIONS

PURPOSE

The purpose of this exercise is to reinforce your understanding of delta and wye connections.

PROCEDURE

1. The three coils in Fig. 8-3 are the phase windings of a three-phase ac generator. The black terminal (dot) on each coil represents the reference end of the coil.

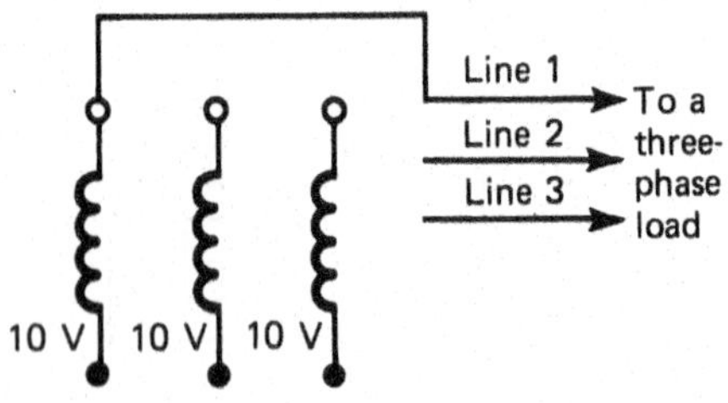

(a)

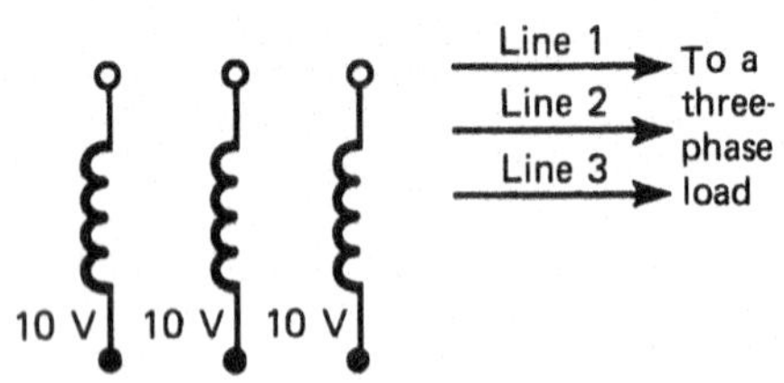

(b)

Fig. 8-3 Making wye and delta connections.

   a. Connect the windings in Fig. 8-3(*a*) in a wye configuration. (The line 1 connection has been made as an example.)

   b. What is the value of the voltage between line 1 and line 2? ______________

   c. What is the value of the voltage between line 1 and line 3? ______________

   d. What is the value of the voltage between the star connection and any of the three lines? ______________

2. Refer to Fig. 8-3(*b*).

   a. Connect these phase windings in a delta configuration.

   b. What is the line voltage of this generator? ______________

NAME ______________________ DATE ______________

8-3 LAB EXPERIMENT: VIEWING AND MEASURING WAVEFORMS

PURPOSE

This experiment will acquaint you with the use of an oscilloscope to view and measure waveforms. It will also introduce you to the audio signal generator. Finally, it will allow you to compare the voltage-measuring capabilities of the DMM and the oscilloscope.

MATERIALS

Qty.		Qty.	
1	oscilloscope with calibrated vertical input and time base	1	signal generator, sinusoidal output, audio range
1	DMM		

INTRODUCTION

An oscilloscope displays a voltage waveform when properly adjusted. That is, it plots a graph with voltage amplitude on the vertical axis and time on the horizontal axis. All oscilloscopes perform this same basic function. However, the details and specific names of the operating controls vary from one oscilloscope to another. An example of an oscilloscope is given in Fig. 8-4. Any oscilloscope can be used with this laboratory experiment including a more modern menu-driven oscilloscope.

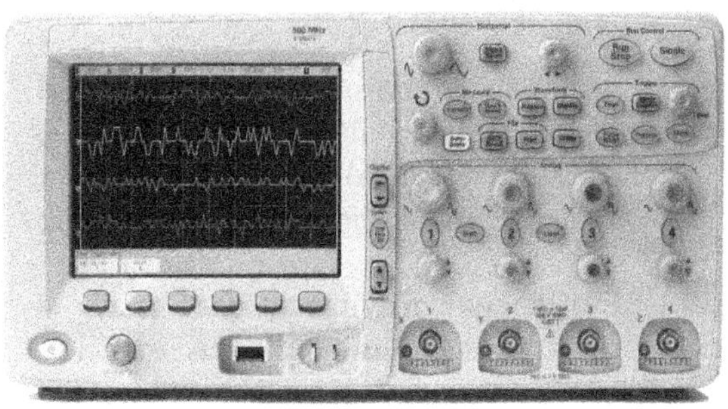

Fig. 8-4 Oscilloscope. (*Courtesy of Tektronix, Inc.*)

The common operating controls, and their functions, for a typical analog oscilloscope are listed and discussed below. After your instructor has demonstrated the operation of your oscilloscope, you can refer to this list to refresh your memory. More details about oscilloscopes can be found in Appendix I of your textbook.

Oscilloscope Operating Controls

INTENSITY Controls the brightness of the waveform. Keep the brightness as low as possible while still displaying a waveform that is easily viewed.

c. Increase the frequency of the generator to 100 kHz (100,000 Hz). Adjust its output so that it again indicates 10 V peak to peak on the oscilloscope. Now measure the output of the generator with the DMM. Does the DMM indicate the same voltage as it did in step 2a?

____

d. Change the generator frequency back to 200 Hz. Set the function switch to the square wave position. Adjust the output voltage to 10 V peak to peak (as indicated on the oscilloscope). Now measure the output voltage of the generator with the DMM. Does it indicate the same voltage as it did in step 2a? ______________ Why?

____

____

3. Turn off the signal generator and the oscilloscope. This section will test your ability to interpret waveforms displayed on an oscilloscope. Refer to Fig. 8-5.

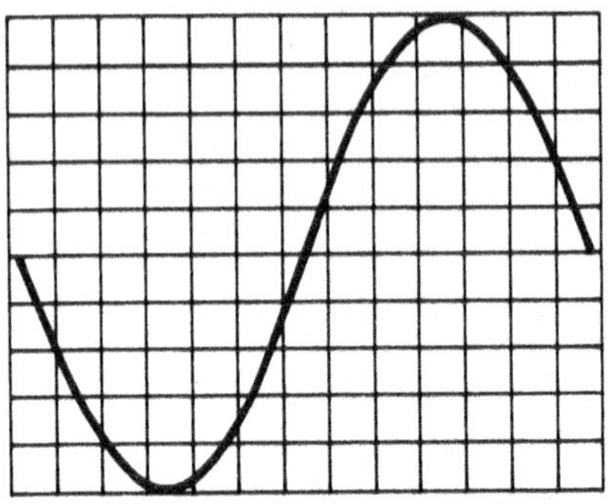

Oscilloscope controls are set for
5 V/div and 0.5 ms/div

Fig. 8-5 Oscilloscope display for experiment 8-3, step 3.

a. What is the period of the waveform? ______________

b. What is the frequency of the waveform? ______________

c. What is the peak voltage of the waveform? ______________

d. How much voltage would the DMM indicate when measuring this waveform? ______________

NAME ______________________________ DATE ______________

8-4 MULTISIM: FREQUENCY AND VOLTAGE

PURPOSE

This Multisim experiment will provide more experience with making measurements with an oscilloscope. Also, it will reemphasize the importance of specifying the correct amplitude values of voltage and current.

MATERIALS

Qty.

Multisim
Multisim file for Fig. 8-6.

PROCEDURE

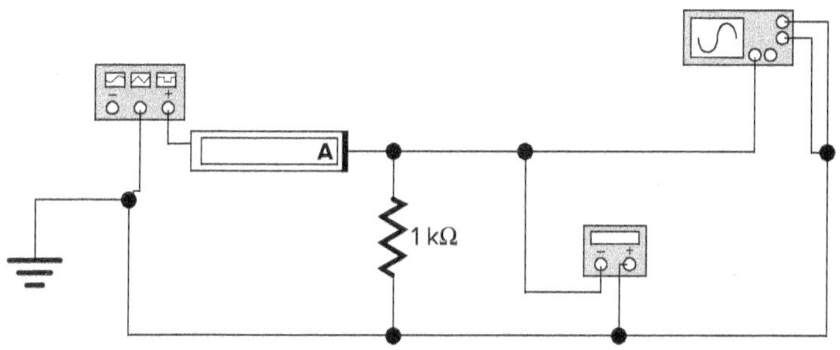

Fig. 8-6 Multisim circuit for experiment 8-4.

1. After opening the Fig. 8-6 Multisim file, activate the circuit.
   - **a.** Using the oscilloscope display, determine and record the period of the waveform. ______________
   - **b.** Calculate and record the frequency of the waveform. ______________
   - **c.** Does the calculated frequency agree with the value specified on the function generator? ______________
   - **d.** Record the voltage measured with the oscilloscope. ______________
   - **e.** Using the voltage measured with the oscilloscope and the specified value of *R*, calculate and record the current. ______________
   - **f.** Does the calculated current agree with the measured current? ______________ Should it? ______________
   - **g.** Which voltage ($V_{p\text{-}p}$, V_p, V_{av}, or V) is specified on the function generator? ______________
   - **h.** Which voltage ($V_{p\text{-}p}$, V_p, V_{av}, or V) is specified on the oscilloscope? ______________
   - **i.** Which voltage ($V_{p\text{-}p}$, V_p, V_{av}, or V) is specified on the multimeter? ______________
   - **j.** Does the ammeter indicate I_{rms}? ______________
   - **k.** Would the oscilloscope frequency or amplitude indications change if the function of the generator were changed to square wave? ______________
   - **l.** Would the voltage indicated by the multimeter change if the function of the generator were changed to square wave? ______________
   - **m.** Check your answers to steps k and l by clicking on the square wave shown on the generator.

8-5 MULTISIM: OSCILLOSCOPE MEASUREMENTS

PURPOSE

This Multisim experiment will give you more practice in using the oscilloscope to gather data. You can use this data to calculate other quantities in the circuit.

MATERIALS

Qty.

Multisim
Multisim file for Fig. 8-7

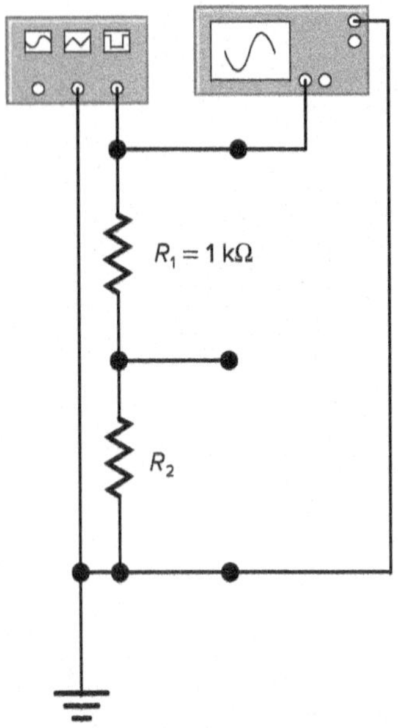

Fig. 8-7 Multisim circuit for experiment 8-5.

PROCEDURE

1. Open the Fig. 8-7 Multisim file and activate the circuit. This circuit is restricted so that you have access only to the oscilloscope to make measurements. Adjust the oscilloscope until you get a two- to three-cycle display that is not clipped.
   a. Measure and record the period of the waveform. ______________
   b. Calculate the frequency of the waveform. ______________
   c. Double-click on the generator symbol. Was your answer to step b correct? ______________
   d. Close the generator display. Now measure and record V_T.
   ____
   e. Next, measure and record the voltage across R_1. ______________
   f. Find, measure, and record the voltage across R_2. ______________
   g. Using your measured values, calculate and record
      (1) The total current ______________
      (2) The resistance of R_2 ______________
      (3) The total power dissipated ______________
      (4) The power dissipated by R_1 ______________
   h. What is I_{av} for this circuit? ______________

NAME ____ DATE ______________

8-6 MULTISIM: HALF-WAVE RECTIFICATION

PURPOSE

Completing this Multisim experiment will help you develop a better understanding of half-wave rectification. It will also give you more experience in using an oscilloscope. And finally, it will provide practice in converting between average, peak, and effective values of ac.

MATERIALS

Qty.

Multisim
Multisim file for Fig. 8-9

INTRODUCTION

Figure 8-8 illustrates how to estimate the dc output of a half-wave rectifier circuit. We know the output consists of sinusoidal pulses equal to the peak value of the ac input. (This assumes an ideal diode that drops no voltage while conducting. A real diode drops approximately 0.7 V while conducting. Unless the voltage to be rectified is comparatively small, we can assume an ideal diode.) The sinusoidal pulses in Fig. 8-8(*a*) can be converted to average-value pulses, as shown in Fig. 8-8(*b*), by using the formula $V_{av} = V_{peak} \times 0.637$. Finally, as shown in Fig. 8-8(*c*), the dc output is equal to half the value of average pulses in Fig. 8-8(*b*) because the duration of each pulse is half of the duration of one cycle of ac input.

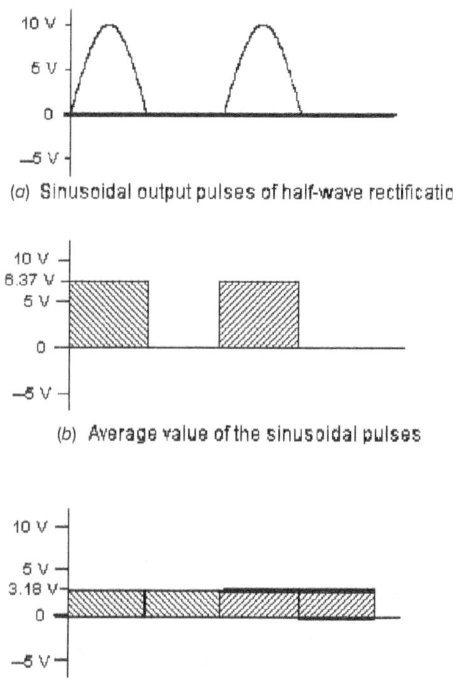

(*a*) Sinusoidal output pulses of half-wave rectificatio

(*b*) Average value of the sinusoidal pulses

(*c*) Average of the two pulses ir(*b*)) is the dc outp

Fig. 8-8 Determining the dc output of half-wave rectification.

PROCEDURE

1. What voltage would you expect the dc meter in Fig. 8-9 to indicate if the generator produces 40 V of ac voltage? ______________
2. Under conditions specified in step 1, what type of waveform should the oscilloscope display? ______________
3. What should be the value of the voltage displayed on the oscilloscope under the conditions given in step 1? ______________
4. The Multisim file for Fig. 8-9 is restricted to the use of the two instruments shown in Fig. 8-9. Open the file and activate the circuit.
   a. What is the value of the ac source voltage? ______________
   b. What is the frequency of the ac voltage? ______________
   c. Is the width of a pulse equal to the width of space between pulses? ______________
   d. Using the data obtained with the oscilloscope, calculate the expected dc output voltage. Show your calculation in the margin of this page. ______________________________
   e. Does your calculated voltage agree within ±10 percent of the voltage indicated by the meter? ______________

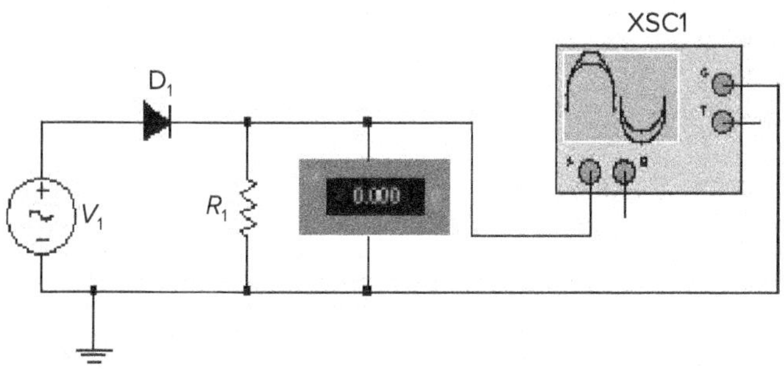

Fig. 8-9 Multisim circuit for experiment 8-6.

8-7 MULTISIM: THREE-PHASE AC CIRCUIT

PURPOSE

Simulating this three-phase circuit with Multisim will reinforce your understanding of a wye-connected, three-phase circuit with balanced and unbalanced loads with and without a fourth (neutral) wire.

MATERIALS

Qty.

Multisim
Multisim file for Fig. 8-10

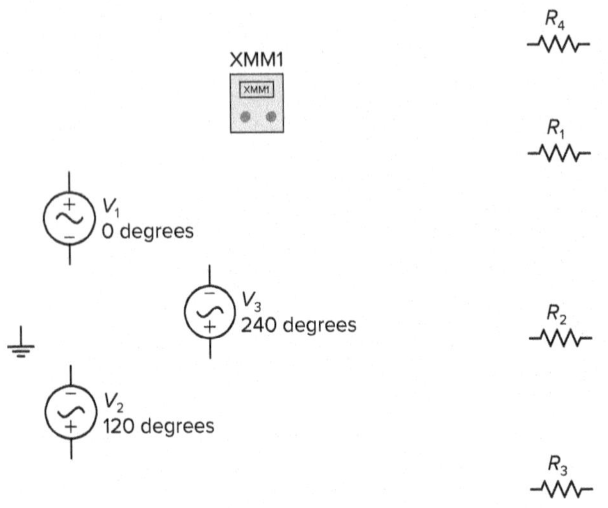

Fig. 8-10 Multisim components for experiment 8-7.

PROCEDURE

1. After opening the Fig. 8-10 Multisim file, use the components provided to construct a three-phase circuit with a wye-connected source and a wye-connected load without a fourth (neutral) wire. Notice that R_4 is not used at this time.
2. For subscript identification, the line connected to R_1 will be line 1, the line to R_2 will be line 2, and the line to R_3 will be line 3.
3. Activate the circuit and measure and record the following

    a. $V_{\text{phase 1}}$ ____________

    b. $V_{\text{phase 2}}$ ____________

    c. $V_{\text{phase 3}}$ ____________

    d. $V_{\text{line 1 to 2}}$ ____________

    e. $V_{\text{line 1 to 3}}$ ____________

    f. $V_{\text{line 2 to 3}}$ ____________

    g. $I_{\text{line 1}}$ ____________

    h. $I_{\text{line 2}}$ ____________

    i. $I_{\text{line 3}}$ ____________

j. V_{R_1} ______________

k. V_{R_2} ______________

l. V_{R_3} ______________

4. Is the load balanced? ______________
5. Calculate and record the value of R_1. ______________
6. Should any voltage exist between the load star point and the source star point? ______________ Check your answer with the voltmeter!
7. Using your measured data, calculate and record the relationship between V_{line} and V_{phase}. ______________
   Is V_{line} about 1.732 times as large as V_{phase}? If not, something is wrong with your circuit.
8. Connect R_4 in parallel with R_1 and then measure and record the following.

   a. $V_{phase\ 1}$ ______________

   b. $V_{phase\ 2}$ ______________

   c. $V_{phase\ 3}$ ______________

   d. $V_{line\ 1\ to\ 2}$ ______________

   e. $V_{line\ 1\ to\ 3}$ ______________

   f. $V_{line\ 2\ to\ 3}$ ______________

   g. $I_{line\ 1}$ ______________

   h. $I_{line\ 2}$ ______________

   i. $I_{line\ 3}$ ______________

   j. V_{R_1} ______________

   k. V_{R_2} ______________

   l. V_{R_3} ______________

9. Summarize the changes that occurred when the load was unbalanced by decreasing the resistance in the phase 1 part of the load. ______________
   ____

10. Leave the load unbalanced, but connect a conductor (neutral) between load and source star points. Now measure and record the following

a. $V_{\text{phase 1}}$ ________________

b. $V_{\text{phase 2}}$ ________________

c. $V_{\text{phase 3}}$ ________________

d. $V_{\text{line 1 to 2}}$ ________________

e. $V_{\text{line 1 to 3}}$ ________________

f. $V_{\text{line 2 to 3}}$ ________________

g. $I_{\text{line 1}}$ ________________

h. $I_{\text{line 2}}$ ________________

i. $I_{\text{line 3}}$ ________________

j. V_{R_1} ________________

k. V_{R_2} ________________

l. V_{R_3} ________________

11. Next, measure and record the current in the neutral conductor. ________________

12. Is the neutral current equal to the difference between I_{line1} and I_{line2}? ________________ It should be!

13. Summarize the changes that occurred when switching from a balanced-load, three-wire-wye connection to an unbalanced load, four-wire-wye connection.

NAME ________________________________ DATE ______________

CHAPTER 9

Power in AC Circuits

TEST: POWER IN AC CIRCUITS

For questions 1 to 18, determine whether each statement is true or false.

1. The angle θ will be 45° when the resistance and reactance in a circuit are equal. **1.** ________
2. The current and voltage are 90° out of phase in a purely resistive ac circuit. **2.** ________
3. A load that contains both resistance and reactance uses power. **3.** ________
4. A circuit has reactance when the current and voltage are out of phase. **4.** ________
5. A circuit has no resistance if the phase shift between current and voltage is exactly 90°. **5.** ________
6. Power factor can be between zero and 1. **6.** ________
7. Phasors rotate in a clockwise direction. **7.** ________
8. Most electric motors can be considered inductive devices. **8.** ________
9. A capacitor possesses reactance. **9.** ________
10. Apparent power can be measured with a wattmeter. **10.** ________
11. The current leads the voltage in an ac motor. **11.** ________
12. All current phasors in a phasor diagram must use the same scale. **12.** ________
13. In Fig. 9-1(*a*) the current is lagging the voltage. **13.** ________
14. In Fig. 9-1(*b*) the voltage is leading the current. **14.** ________
15. True power is always less than apparent power in circuits that have some reactance. **15.** ________
16. The power factor is equal to the sine of angle θ. **16.** ________
17. The angle between the total voltage and the total current is called θ. **17.** ________
18. When the power factor is 1, the most efficient use of distribution lines and equipment occurs. **18.** ________

For questions 19 to 22, choose the letter that best completes each statement.

19. Unless otherwise stated, ac power is specified in **19.** ________
    - **a.** Instantaneous value
    - **b.** Average value
    - **c.** Effective value
    - **d.** Peak value
20. The form of opposition to alternating current that uses no power is called **20.** ________
    - **a.** Reluctance
    - **b.** Reactance
    - **c.** Resistance
    - **d.** Resistivity

NAME ____ DATE ____________________

9-1 LAB EXPERIMENT: TRUE POWER AND APPARENT POWER

PURPOSE

This experiment will provide experience in measuring true and apparent power in several types of loads. You will gain experience in using an ac ammeter, voltmeter, and wattmeter. The results of this activity will verify that the power in an inductive circuit is not equal to current times voltage.

MATERIALS

Qty.		Qty.	
1	ac ammeter, 10-A range	1	heater or hair dryer, 500 W
1	ac wattmeter, 1000-W range, 300 V, 7 A	1	ac motor, ⅓ or ½ hp
1	ac voltmeter, 150-V range	1	Isolated variable transformer, 0- to 120-V ac, 7 A

CAUTION **The circuits in this experiment are powered from a 120-V ac outlet. A shock from such a circuit can be fatal. Make all connections in the circuit before connecting the variable transformer to the 120-V outlet. Do not touch any part of the circuit after it is connected to the 120-V source.**

PROCEDURE

1. Connect the circuit shown in Fig. 9-3(*a*). Be sure to use an isolated variable transformer. Notice that the ammeter and the voltmeter have no polarity markings because ac changes polarity every half cycle. Set the variable transformer control for minimum output voltage. Then plug the transformer into a 120-V outlet. Now, rotate the variable transformer control until the voltmeter indicates 120 V.

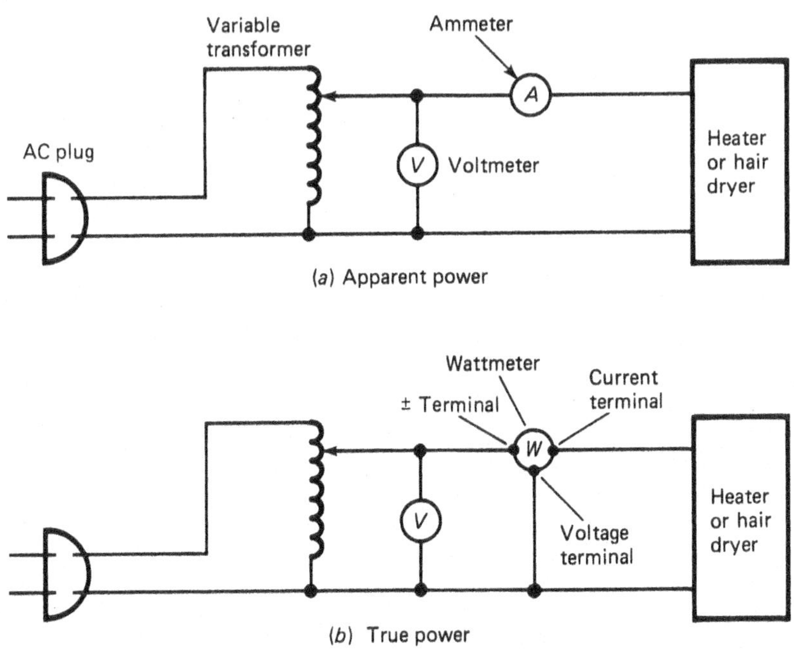

Fig. 9-3 Measuring the power of the hair dryer.

a. Read and record the current indicated by the ammeter. _____________

b. Return the variable transformer to minimum setting and unplug it from the 120-V outlet. Using the measured current and voltage, calculate and record the apparent power of the heater. _____________

c. Change the ammeter of Fig. 9-3(*a*) to a wattmeter as shown in Fig. 9-3(*b*). Notice that the wattmeter symbol shows three leads. However, many wattmeters have four terminals. Two terminals are for the current coil and two for the voltage coil. One terminal of each coil is marked with a ± sign. Connect the two terminals with the ± signs together and treat them as if they were a single terminal. Be sure the current and voltage terminals are connected as shown in Fig. 9-3(*b*). Now plug the transformer in and adjust it for 120 V as indicated on the voltmeter. Read and record the power indicated by the wattmeter.

____

d. Reset the transformer for minimum voltage and unplug it. Compare the apparent power and the true power for this resistive load. Are they the same? _____________ Why?

____

____

e. What is the power factor of this circuit? _____________

2. Connect the circuit of Fig. 9-4(*a*). Check to be sure the variable transformer is set for minimum voltage. Plug in the transformer and slowly increase the voltage while watching the ammeter. Increase the voltage slowly so that the ammeter pointer moves smoothly. Continue increasing the voltage until the voltmeter indicates 120 V.

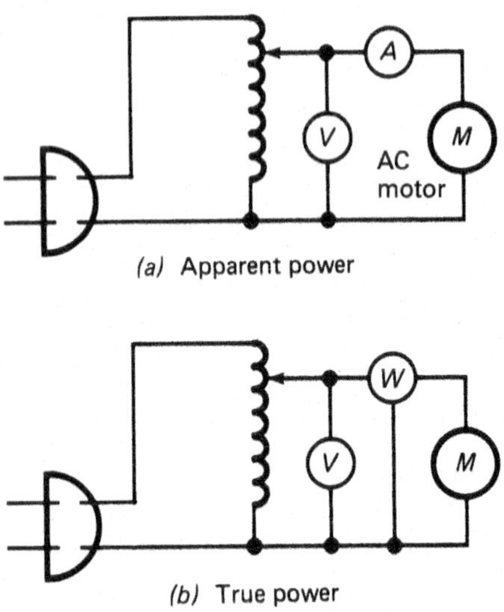

Fig. 9-4 Measuring the input to a motor.

a. Read and record the current. _____________

b. Reduce the voltage to minimum value and unplug the transformer. Calculate and record the apparent power. _____________

c. Connect the circuit of Fig. 9-4(*b*). Again, be sure the variable transformer control is set for minimum voltage before plugging it into the 120-V outlet. Now increase the transformer output voltage slowly so as not to damage the wattmeter. Adjust the transformer for 120 V. Read and record the power. _____________

d. Again reduce the voltage to minimum and unplug the transformer. Are the apparent power and the true power the same for an ac motor? ______________ Why?

____

____

e. Determine and record the power factor for this circuit. ______________

f. Determine and record angle θ for this circuit. ______________

9-2 ADVANCED PROBLEMS

9-1. Refer to Fig. 9-5 and answer the following:

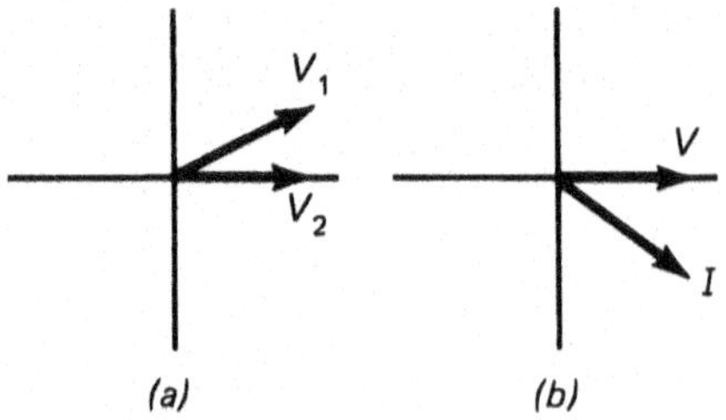

Fig. 9-5 Phasors.

a. In Fig. 9-5(*b*), which phasor is leading? ______________
b. In Fig. 9-5(*a*), must V_2 be numerically smaller than V_1? ______________
c. In Fig. 9-5(*b*), is *I* leading or lagging *V*? ______________
d. In Fig. 9-5(*b*), must *V* be numerically larger than *I*? ______________
e. Are the phasors in Fig. 9-5(*b*) representing an inductive or a capacitive circuit? ______________

9-2. A circuit has 4 V of resistive voltage and 3 V of inductive voltage. In the margin of this page, draw a phasor diagram showing the total voltage. Use a scale of 1 cm = 1 V.

9-3. A motor draws 7 A from a 240-V source at a power factor of 70 percent. Find the following:
a. The apparent power ______________
b. The actual (true) power ______________
c. Angle θ ______________
d. The resistive current ______________
e. The reactive current ______________

9-4. The apparent power in a circuit is 690 W and the actual power is 550 W. The source voltage is 115 V. Find the following:
a. Total current ______________
b. Cosine θ ______________
c. Angle θ ______________
d. Resistive current ______________

NAME ______________________________ DATE ______________

9-3 MULTISIM: PHASE SHIFT

PURPOSE

As you vary the relative amounts of resistive and reactive loads in a circuit, the phase angle between *I* and *V* will vary. This Multisim experiment allows you to observe the phase shift between the current and voltage waveforms as the combination load is varied.

MATERIALS

Qty.

Multisim
Multisim file for Fig. 9-6

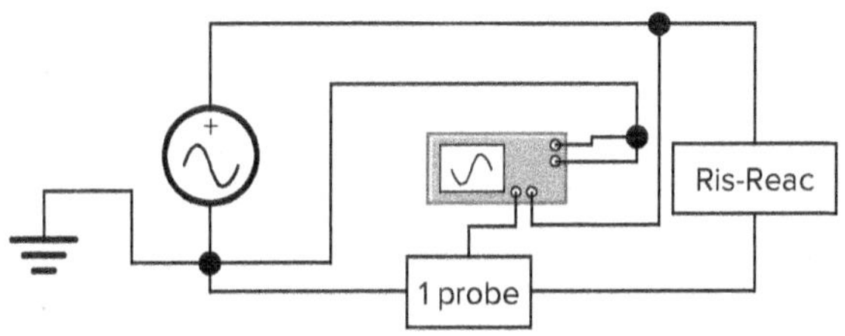

Fig. 9-6 Multisim circuit for experiment 9-3.

PROCEDURE

1. After opening the Multisim file for Fig. 9-6, activate the circuit.
   a. Click the pause button to lock the waveforms. (You may have to click the button several times to get a locked waveform.) Measure and record the phase shift. ______________
   b. Restart the circuit and make the load more resistive. Does this increase or decrease angle theta? ______________
   c. Measure and record the minimum angle theta available with this load. ______________
   d. Measure and record the maximum angle theta available with this load. ______________

9-4 MULTISIM: THETA AND POWER FACTOR

PURPOSE

This Multisim experiment provides practice in measuring ac voltages and current to be used in determining other circuit quantities.

MATERIALS

Qty.

Multisim
Multisim file for Fig. 9-7

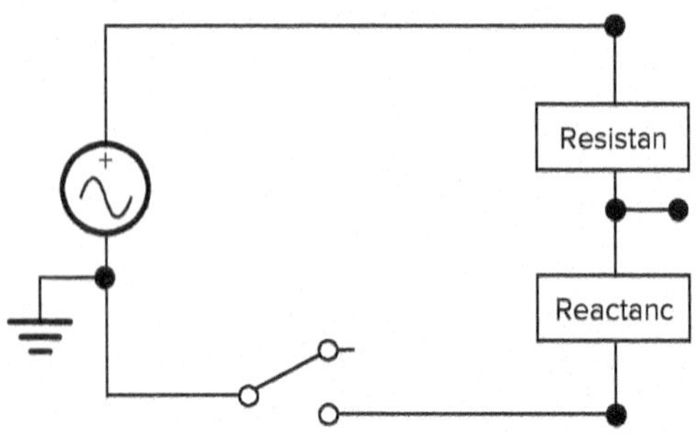

Fig. 9-7 Multisim circuit for experiment 9-4.

PROCEDURE

1. After opening the Multisim file for Fig. 9-7, activate the circuit.
   - **a.** Measure and record the voltage across the resistive part of the circuit.
     ____
   - **b.** Measure and record the voltage across the reactive part of the circuit.
     ____
   - **c.** From the results obtained in steps a and b, is theta greater than or less than 45°? ______________
   - **d.** Measure and record the total voltage and the total current.
     ______________ ______________
   - **e.** Using your measured data, calculate and record the following:

     $P =$ ______________

     $P_{app} =$ ______________

     $\theta =$ ______________

     PF = ______________

NAME ______________________________ DATE ______________

9-5 MULTISIM: THREE-PHASE POWER

PURPOSE

Simulating the three-phase circuit in Fig. 9-8 using Multisim will give you experience in measuring power in a three-phase circuit.

MATERIALS

Qty.

Multisim
Multisim file for Fig. 9-8

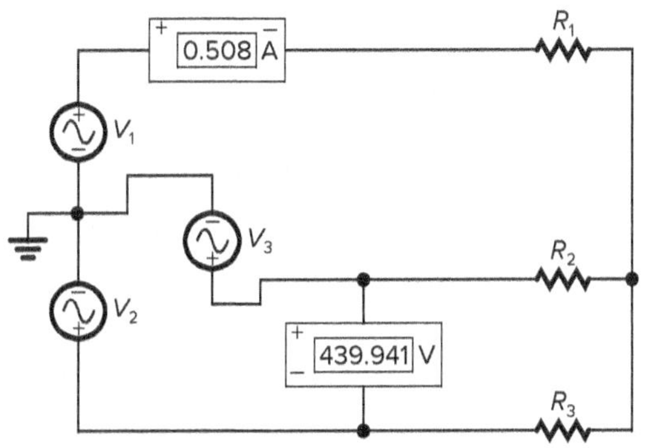

Fig. 9-8 Multisim circuit for experiment 9-5.

PROCEDURE

1. What type of connection is used for the three-phase source in Fig. 9-8? ______________
2. What type of connection is used for the load in Fig. 9-8? ______________
3. What is the power factor in Fig. 9-8? ______________
4. The three resistors in Fig. 9-8 have the same resistance. Knowing this and the information given in Fig. 9-8, calculate and record the value of:

   a. V_{phase} ______________

   b. P_{phase} ______________

   c. R_{phase} ______________

5. Write the formula(s) you used in calculating the values in

   a. step 4a ______________

   b. step 4b ______________

   c. step 4c ______________

6. Open and activate the Fig. 9-8 Multisim file.
   a. Measure and record the phase voltage. ______________
   b. Calculate and record the phase power using the measured values of phase voltage and phase current. ______________
   c. Measure and record the value of any one of the resistors. ______________
7. Do your answers in step 4 agree with your answers in step 6? ______________ They should!
8. Double-click on the XWM1 icon. This instrument is a combination wattmeter and power-factor meter. Notice that it has four terminals. If you connect the voltage negative terminal to the current positive terminal and consider this connection as a single terminal, then you can treat it like the three-terminal wattmeter shown in Fig. 9-3(*b*). Using the wattmeter, measure and record the phase power and power factor of the circuit. ______________ ______________
9. Do your answers in step 3 and step 4b agree with your answers in step 8? ______________ They should!

NAME ______________________________ DATE ______________

CHAPTER 10

Capacitance

TEST: CAPACITANCE

For questions 1 to 24, determine whether each statement is true or false.

1. Insulating material is used to make the dielectric in a capacitor. **1.** ________

2. Electrons readily flow through the dielectric of a capacitor. **2.** ________

3. The voltage across the plates of a capacitor decreases as energy is stored in the capacitor. **3.** ________

4. For its weight and size, a capacitor stores more energy than a battery. **4.** ________

5. Air has a larger dielectric constant than mica does. **5.** ________

6. All capacitors have a positive temperature coefficient. **6.** ________

7. Doubling the plate area of a capacitor doubles its capacitance if all other factors remain unchanged. **7.** ________

8. For a given size and weight, an electrolytic capacitor has more capacitance than a ceramic capacitor. **8.** ________

9. The outside foil of a tubular capacitor is identified by the curved line on a capacitor symbol. **9.** ________

10. Most foil capacitors are self-healing. **10.** ________

11. Ceramic capacitors are often used to filter pulsating direct current. **11.** ________

12. Bypass capacitors are used in electronic photoflash equipment. **12.** ________

13. The color band on the end of a film capacitor identifies the outside foil. **13.** ________

14. Direct current continuously flows through a capacitor. **14.** ________

15. The voltage leads the current by 90° in an ideal capacitor. **15.** ________

16. A nonelectrolytic capacitor usually has more than 1 percent energy loss. **16.** ________

17. Capacitance uses no power in either an ac or a dc circuit. **17.** ________

18. Capacitors block alternating current in a circuit. **18.** ________

19. A capacitor returns energy to the main energy source twice each cycle. **19.** ________

20. The quality of a capacitor can be indicated by its power factor. **20.** ________

21. The larger of two series capacitors in an ac circuit drops more voltage. **21.** ________

22. The larger of two parallel capacitors in an ac circuit passes more current. **22.** ________

23. A capacitor changes its voltage by 36.8 percent during one time constant. **23.** ________

24. A ceramic capacitor that is not defective should indicate about midscale on the highest range of an ohmmeter. **24.** ________

For questions 25 to 32, choose the letter that best completes each statement.

25. The dielectric material of a capacitor **25.** ________

a. Seals the capacitor from moisture
b. Is a conductor
c. Is an insulator
d. Connects to the leads

26. Capacitors have a voltage rating because the 26. ____________
   a. Plates can store only a limited number of electrons
   b. Dielectric can store only a limited number of electrons
   c. Dielectric can withstand the force of only a limited electric field
   d. Plates can withstand the force of only a limited electric field
27. The voltage rating of a capacitor is referred to as its 27. ____________
   a. ESR rating
   b. DCWV rating
   c. DFV rating
   d. PFV rating
28. The capacitance of a capacitor is not determined by the 28. ____________
   a. Thickness of the dielectric
   b. Temperature
   c. Plate area
   d. Frequency
29. The dielectric constant of a material indicates its ability to 29. ____________
   a. Resist current
   b. Conduct current
   c. Store energy
   d. Withstand voltage
30. A film capacitor could have a dielectric composed of 30. ____________
   a. Polystyrene
   b. Ceramic
   c. Mica
   d. Metal oxide
31. If two capacitors of unequal value are series-connected to a dc source, 31. ____________
   a. Each capacitor will develop all the source voltage
   b. Each capacitor will develop half the source voltage
   c. The smaller capacitor will develop more voltage
   d. The larger capacitor will develop more voltage
32. Minimum specifications for ordering ceramic capacitors need not include 32. ____________
   a. Voltage rating
   b. Current rating
   c. Tolerance
   d. Capacitance

For questions 33 to 44, supply the missing word or phrase in each statement.

33. A(n) ____________ capacitor is polarized. 33. ____________
34. The opposition of a capacitor is called ____________. 34. ____________
35. A capacitor is over 99 percent charged after ____________ time constants. 35. ____________
36. A capacitor's opposition is ____________ proportional to frequency. 36. ____________
37. ____________ capacitors generally have a lower Q than other types of capacitors. 37. ____________
38. The dielectric in an electrolytic capacitor is ____________. 38. ____________
39. The symbol for capacitive reactance is ____________. 39. ____________
40. The base unit of capacitance is the ____________. 40. ____________
41. The abbreviation or symbol for capacitance is ____________. 41. ____________
42. Physical devices that store energy in an electric field are called ____________. 42. ____________
43. In terms of other base units, the base unit of capacitance equals a(n) ____________. 43. ____________
44. Aluminum or ____________ is used for the plates of an electrolytic capacitor. 44. ____________

NAME ______________________________ DATE ______________

For questions 45 to 54, solve each problem. Be sure to include units in your answer when appropriate.

45. How much energy can a capacitor store if it is rated at 560 μF and 400 V? **45.** ____________

46. How many picofarads are there in 0.0015 μF? **46.** ____________

47. What is the reactance of a 0.22-μF, 250-V capacitor at 400 Hz? **47.** ____________

48. What is the equivalent capacitance of a 220-pF capacitor and a 100-pF capacitor connected in parallel? **48.** ____________

49. What is the equivalent capacitance of the capacitors in question 48 when connected in series? **49.** ____________

50. What is the time constant of a 470-kΩ resistor and a 0.33-μF capacitor series-connected to a 40-V dc source? **50.** ____________

51. For the circuit in Fig. 10-1, determine the following: **51.** *Place answers below.*

a. The voltage across the 0.8-μF capacitor **a.** ____________

b. The total opposition **b.** ____________

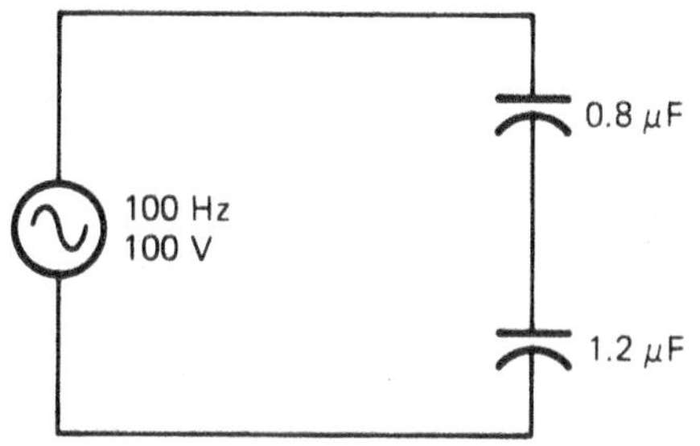

Fig. 10-1 Circuit for question 51.

52. For the circuit in Fig. 10-2, determine the following: **52.** *Place answers below.*

a. The current through the 0.001-μF capacitor **a.** ____________

b. The total current **b.** ____________

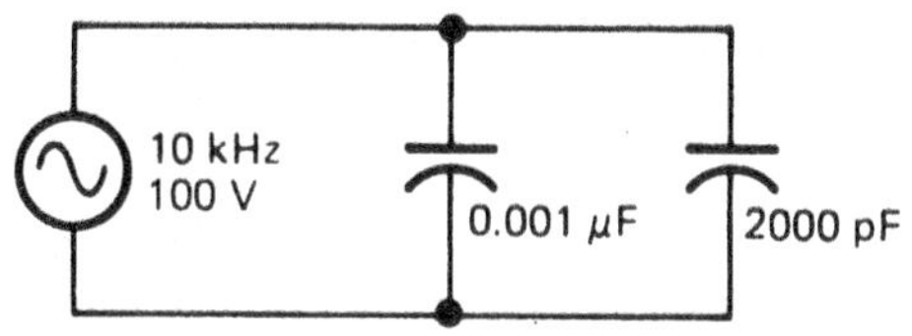

Fig. 10-2 Circuit for question 52.

53. Express the value of a 33,000-pF capacitor in nanofarads. **53.** ____________

54. For the circuit in Fig. 10-3, determine: **54.** *Place answers below.*

a. I_{C2} **a.** ____________

b. I_{C3} **b.** ____________

c. C_T **c.** ____________

d. X_{CT} **d.** ____________

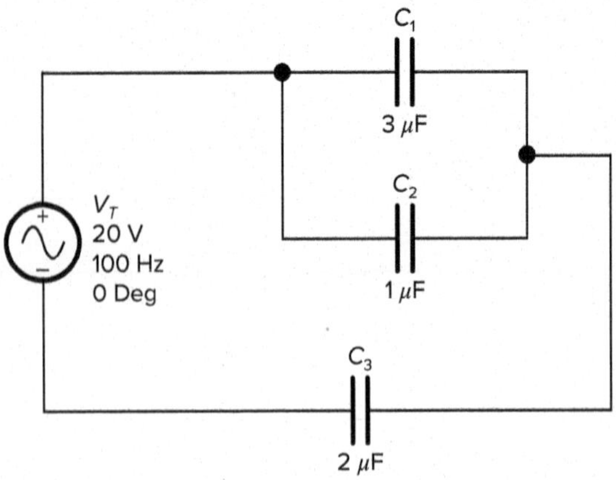

Fig. 10-3 Circuit for question 54.

NAME ______________________ DATE ______________

10-1 LAB EXPERIMENT: CAPACITORS IN AC AND DC CIRCUITS

PURPOSE

This experiment will provide visual evidence of the major characteristics of capacitors in ac and dc circuits. It will also provide experience in correctly connecting electrolytic capacitors in ac and dc circuits.

MATERIALS

Qty.	
1	power supply, 6 V ac, 60 Hz (a 6-V transformer)
1	power supply, 10 V dc
2	capacitors, electrolytic, 1000 μF, 50 WV dc (minimum)
1	capacitor, nonelectrolytic, 1 μF, 50 WV dc (minimum)
1	capacitor, nonelectrolytic, 0.47 μF, 50 WV dc (minimum)
1	lamp, no. 47 (6.3 V, 0.15 A)
1	holder for no. 47 lamp
1	DMM, 10-MΩ minimum input resistance
1	switch, SPST, 1 A, 125 V

PROCEDURE

1. Construct the circuit of Fig. 10-4. Recheck the circuit to be sure you have the correct polarity on C_1. Open S_1. Discharge C_1 by momentarily connecting a jumper lead from one lead of the capacitor to the other lead. Remove the jumper. Now, while watching the lamp, close S_1.

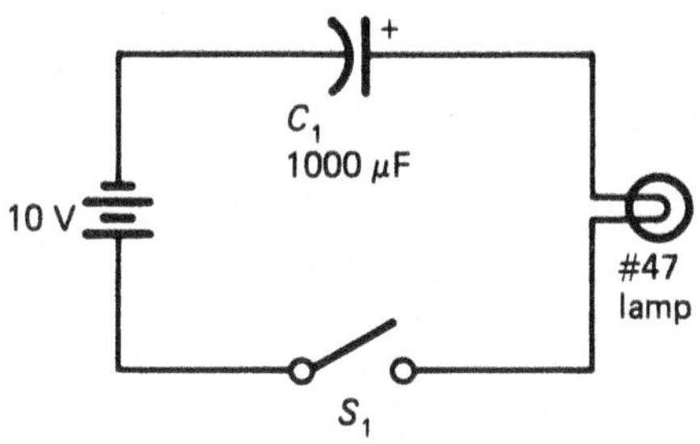

Fig. 10-4 Circuit for experiment 10-1, step 1.

a. Did the lamp flash? ______________

b. Is current still flowing in the circuit? ______________ Why?

____

____

c. Check your answer to step 1b by measuring the voltage drop across the lamp. (Remember that a voltage drop across the lamp indicates a current flow.) What is the voltage across the lamp? ______________

d. Now open S_1. Is the capacitor now charged? If it is, it should provide a voltage even though S_1 is open. ______________

e. What value of voltage should the capacitor provide? ______________

f. Check your answer to step 1e by measuring the voltage across capacitor C_1.

g. Further evidence that a capacitor stores energy can be obtained by rapidly discharging the capacitor. Close S_1 to recharge the capacitor. Open S_1. Now short out the capacitor with a jumper lead. What happened?

____

____

2. The two electrolytic capacitors in Fig. 10-5 are connected positive terminal to positive terminal. This arrangement provides the equivalent of one nonpolarized 500-μF capacitor. Construct the circuit shown in Fig. 10-5 (be sure the capacitors are connected as indicated on the schematic). Close the switch and observe the lamp.

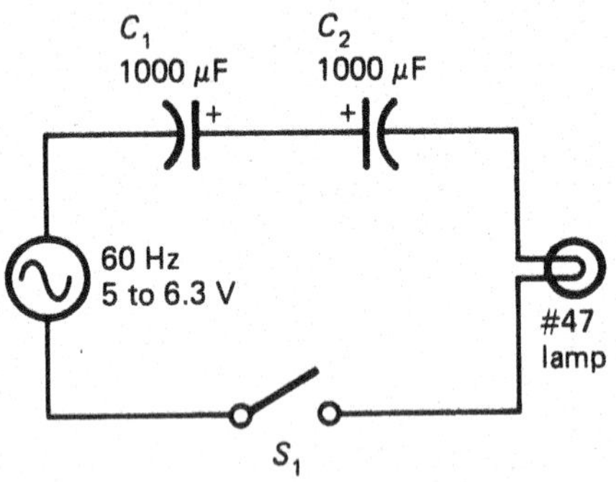

Fig. 10-5 Circuit for experiment 10-1, step 2.

   a. Does the lamp light? ______________

   b. Will the lamp light at this brightness for an extended period of time? ______________

   c. Measure and record the voltage across the lamp. ______________

   d. Measure and record the voltage across the combination of C_1 and C_2 (negative lead of C_1 to negative lead of C_2). ______________

   e. Does the lamp or the capacitor offer more opposition to the alternating current? ______________

3. Construct the circuit shown in Fig. 10-6. Open S_1 and discharge C_1 with a jumper lead. Then discharge C_2 with the jumper lead (and remove the jumper lead). This procedure ensures that both capacitors are fully discharged before power is applied to the circuit. Now close S_1 and allow the capacitors to charge.

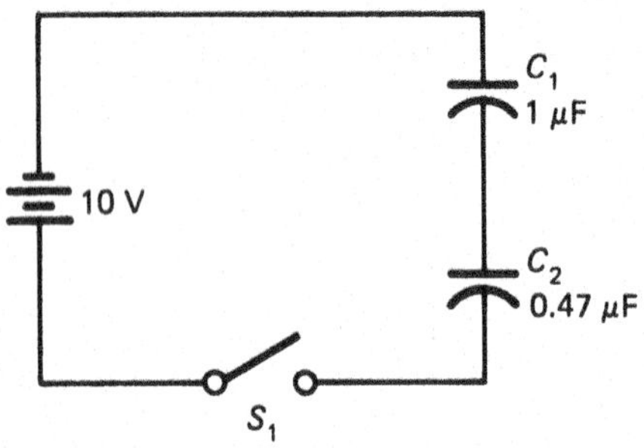

Fig. 10-6 Circuit for experiment 10-1, step 3.

   a. How much voltage should be across C_1? ______________

   b. How much voltage should be across C_2? Open S_1. ______________

   In making the following measurements, be prepared to read your voltmeter as quickly as possible. Quick reading is necessary to achieve accurate results because the capacitor's voltage will be discharged through the meter.

   c. Measure and record the voltage across C_1. ______________

   d. Does your measured voltage agree (within 10 percent) with the voltage you predicted in step 3a? ______________

e. Measure the voltage across C_2 and compare it with the value predicted in step 3b. Do the two values agree within 10 percent of each other? ______________

4. As a review, answer the following questions:
   a. Does current flow in a dc circuit while a capacitor is charging?
   b. Does current flow in a dc circuit after a capacitor has charged?
   c. Does a capacitor block alternating current? ______________
   d. Does a capacitor block direct current? ______________
   e. How does dc voltage distribute in a series capacitor circuit?
   ____
   ____

NAME ______________________ DATE ______________

10-2 LAB EXPERIMENT: RESISTOR-CAPACITOR TIME CONSTANTS

PURPOSE

The purpose of this experiment is to measure *RC* time constants of both charging and discharging circuits. These measurements will verify that $T = RC$.

MATERIALS

Qty.

1 power supply, 10 V dc
1 capacitor, nonelectrolytic, 1 μF $\pm$10%, 50 WV dc (minimum)
1 capacitor, nonelectrolytic, 0.47 μF $\pm$10%, 50 WV dc (minimum)
1 switch, SPST, 1 A, 125 V ac
1 DMM, 11-MΩ input resistance

INTRODUCTION

In this experiment we will be using the internal resistance of the voltmeter as the resistance in our *RC* circuits. Thus, the meter will serve a dual function. First, it serves as the resistor in the circuit, and second, it measures the voltage across itself [Fig. 10-7(*a*)]. The net effect is like having a voltmeter with infinite input resistance measuring the voltage across an 11-MΩ resistor, as in Fig. 10-7(*b*).

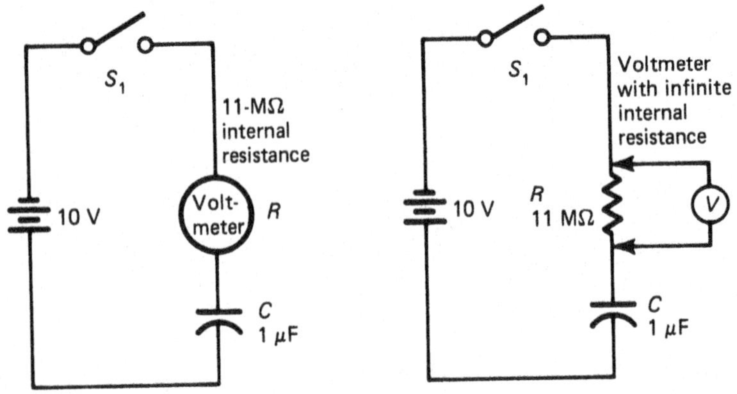

(*a*) The circuit (*b*) Equivalent circuit

Fig. 10-7 Circuit for experiment 10-2, step 1.

In Fig. 10-7(*a*) we will be measuring the voltage across *R*. However, we really want to know the voltage across *C*. Then we will know when it has charged to 63.2 percent of the available voltage (one time constant). This is really no problem. Since we know V_T and are measuring V_R, we can use Kirchhoff's voltage law to find V_C:

$$V_T = V_R + V_C$$

Therefore,

$$V_C = V_T - V_R$$

We can use this last expression to find the instantaneous value of V_C at any instant we know V_R. For instance, if after 11 s the meter indicates 3.7 V, the capacitor will have

$$\text{Voltage across } C = 10\text{ V} - 3.7\text{ V} = 6.3\text{ V}$$

NAME ______ DATE ______

PROCEDURE

1. **a.** How many seconds are there in one time constant for the circuit in Fig. 10-7(*a*)? ______
   b. At the end of one time constant, how many volts should be across the capacitor? ______
   c. At the end of one time constant, how many volts should be across the resistor (meter)? ______
   d. Construct the circuit shown in Fig. 10-7(*a*). Open the switch, and discharge the capacitor by connecting a jumper lead across it. Remove the jumper. Now close the switch, and read the voltmeter at the end of one time constant (11 s). Does the reading agree (within 15 percent) with the value you predicted in step 1c? (If not, repeat your calculations and measurements.) Record your results in the first row of the data table in Fig. 10-8(*a*). ______

Time (seconds)	V_R (volts)	V_C ($V_T - V_R$) (volts)
11		
22		
33		
44		
55		

(*a*)

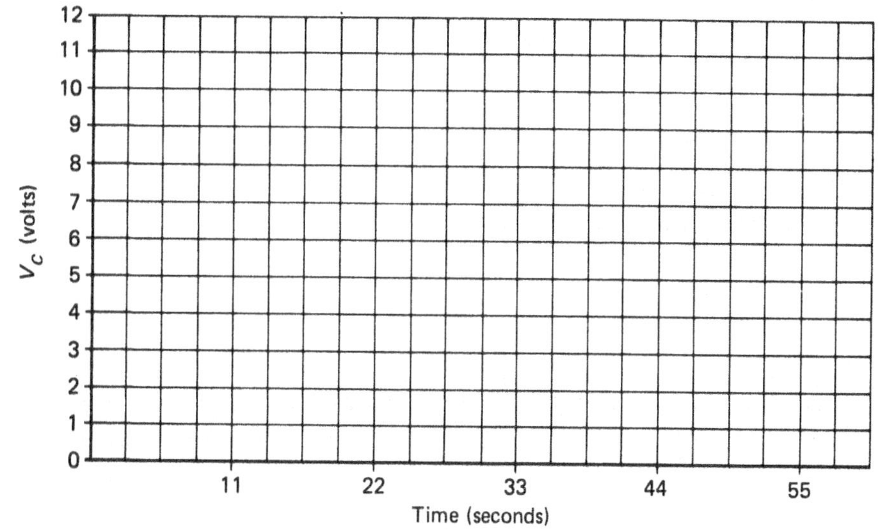

(*b*)

Fig. 10-8 Plotting a charge curve. (*a*) Data table for recording charging voltages. (*b*) Graph paper for plotting a charge curve.

   e. How much voltage would you expect across the resistor at the end of two time constants? ______ Check your answer by opening S_1, discharging *C*, closing S_1, and reading the meter at the end of 22 s. (You may switch ranges on the meter as the capacitor charges to obtain more accurate readings.) Record your results in the second row of the data table. Repeat this process until you have completed the data table in Fig. 10-8(*a*).
   f. Now, using the data collected in Fig. 10-8(*a*), plot five points (time versus V_C) on the graph paper in Fig. 10-8(*b*). Then connect the five points together to form an approximate charge curve for a capacitor. Does this curve resemble the universal charge curve for a capacitor?

   ______

2. When the switch in Fig. 10-9 is closed, the capacitor will charge. When the switch is open, the capacitor will discharge through the resistance of the meter.
   a. What is the time constant of the circuit in Fig. 10-9? ______
   b. One time constant after opening S_1 in Fig. 10-9, how much voltage will the voltmeter indicate? ______

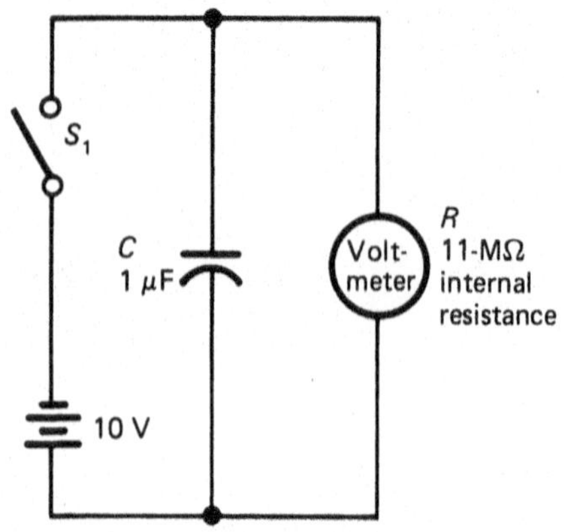

Fig. 10-9 Circuit for experiment 10-2, steps 2 and 3.

c. Construct the circuit of Fig. 10-9 and check your answer to step 2b. Record your answer in the first row of the data table of Fig. 10-10(*a*).

d. Now close S_1 to recharge the capacitor. Open S_1 and measure V_C at the end of two time constants. Record your reading in the data table of Fig. 10-10(*a*). Repeat this procedure for three, four, and five time constants.

Time (seconds)	V_C (volts)
11	
22	
33	
44	
55	

(*a*)

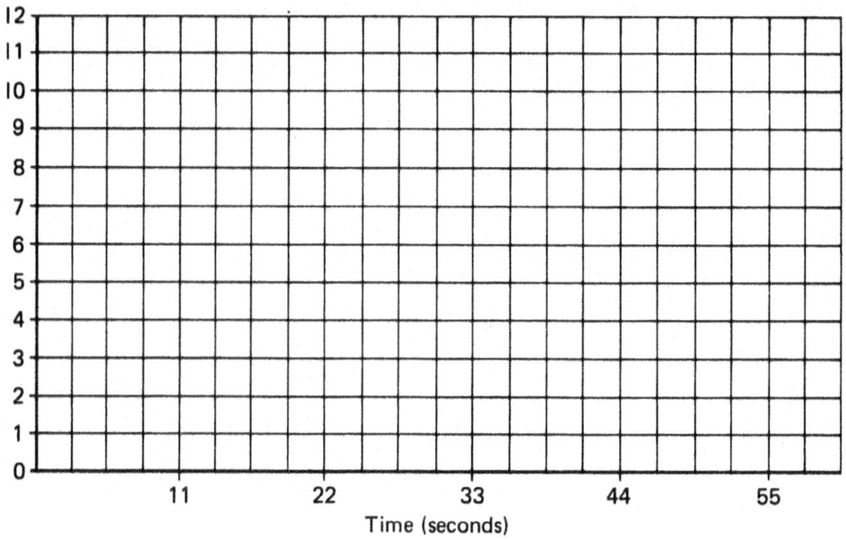

(*b*)

Fig. 10-10 Plotting a discharge curve. (*a*) Data table for recording discharging voltages. (*b*) Graph paper for plotting a discharge curve.

e. Now, use this data to plot and draw a discharge curve on the graph paper in Fig. 10-10(*b*).

f. Is there any difference between the time constant of a charging capacitor and a discharging capacitor? _______________

g. Are the curves you constructed in Figs. 10-8(*b*) and 10-10(*b*) mirror images of each other? _______________

3. a. If the capacitance in Fig. 10-9 were decreased, would the time constant increase or decrease? _______________

b. Check your answer by changing the capacitor in Fig. 10-9 from 1 μF to 0.47 μF. Then measure and record how long it takes the 0.47-μF capacitor to discharge to 3.68 V. _______________

c. Would changing the source voltage in Fig. 10-9 change the time constant of the circuit? _______________

d. Would paralleling the voltmeter with a resistor change the time constant in Fig. 10-9? _______________

NAME ______________________ DATE ______________

10-3 LAB EXPERIMENT: RELATIONSHIPS BETWEEN FREQUENCY, CAPACITANCE, AND REACTANCE

PURPOSE

This experiment will allow you to verify the relationships between *f*, *C*, and X_C. The data collected will lead to the conclusion that X_C is inversely proportional to *C* and *f*. This experiment will also verify the capacitive reactance formula.

MATERIALS

Qty.

1 ammeter, ac, 0 to 5 mA, 148-Ω (or less) internal resistance, ±3% (or less) accuracy
1 DMM or VOM, 10-V ac range
1 capacitor, nonelectrolytic, 1 μF ±10%, 50 WV dc (minimum)
1 capacitor, nonelectrolytic, 0.47 μF ±10%, 50 WV dc (minimum)
1 signal generator, sinusoidal, 20- to 1000-Hz range

PROCEDURE

1. **a.** Construct the circuit shown in Fig. 10-11 using a 1-μF capacitor for C_1. Set the generator frequency for 100 Hz. Adjust the output control for 3 V across the capacitor. Record the current flowing through the capacitor. ______________

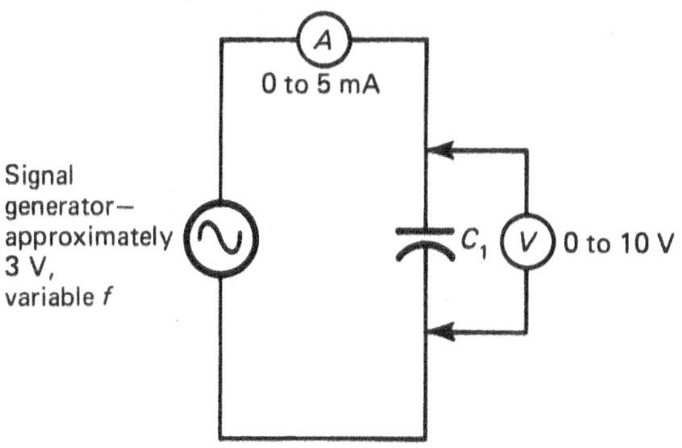

Fig. 10-11 Circuit for experiment 10-3.

b. Using your measured current and voltage, compute and record the reactance of the capacitor using $X_C = V_C/I_C$. ______________

c. Calculate and record the reactance using the reactance formula. ______________

$$X_C = \frac{1}{6.28fC}$$

d. Are the values computed in steps 1b and 1c within 10 percent of each other? (If not, repeat steps 1b and 1c.) ______________

e. Do your results confirm the reactance formula (within the tolerances of the meter, the generator, and the capacitor)? ______________

2. **a.** Suppose the generator frequency is changed to 50 Hz while the voltage remains at 3 V. What will happen to the reactance?

____

____

b. What will happen to the circuit current?

____

____

c. Change the generator frequency to 50 Hz and record the new current. _______________

d. Does your answer to step 2c support your answer to step 2b? _______________

e. How much current would you expect if the frequency were changed to 200 Hz and V_C remained at 3 V? _______________

f. Change the frequency to 200 Hz and adjust the generator for 3 V across the capacitor. Record the current. _______________

g. Is capacitive reactance inversely proportional to frequency? _______________

h. If the frequency were changed to 400 Hz, how much current would you expect through *C*? _______________

3. **a.** Change the capacitor in Fig. 10-11 to 0.47 μF. Leave the frequency at 200 Hz and adjust the output voltage until V_C equals 3 V. How does the new circuit current compare with the current measured in step 2f? _______________ Why?

____

____

b. If the frequency is reduced to 100 Hz and V_C remains at 3 V, how much current should flow in the circuit? Check your answer by reducing the frequency to 100 Hz and measuring the current. _______________

c. What relationship exists between capacitance and capacitive reactance?

____

____

4. As a review, answer the following questions:

a. If both *f* and *C* double, how much will X_C change? _______________

b. If *f* doubles and *C* is reduced to half, how much will X_C change? _______________

c. If both *f* and *C* are reduced to half, how much will X_C change? _______________

10-4 LAB EXPERIMENT: SERIES AND PARALLEL CAPACITORS

PURPOSE

This experiment will illustrate voltage drops and current distribution in series and parallel capacitor ac circuits. You will calculate equivalent capacitance and reactance of series and parallel circuits.

MATERIALS

Qty.

1 ammeter, ac, 0 to 5 mA, 148-Ω (or less) internal resistance, ±3% (or less) accuracy
1 DMM or VOM, 10-V ac range
1 capacitor, nonelectrolytic, 1 μF ±10%, 50 WV dc (minimum)

Qty.

1 capacitor, nonelectrolytic, 0.47 μF ±10%, 50 WV dc (minimum)
1 signal generator, sinusoidal, 20- to 1000-Hz range

PROCEDURE

1. Refer to the series circuit in Fig. 10-12.

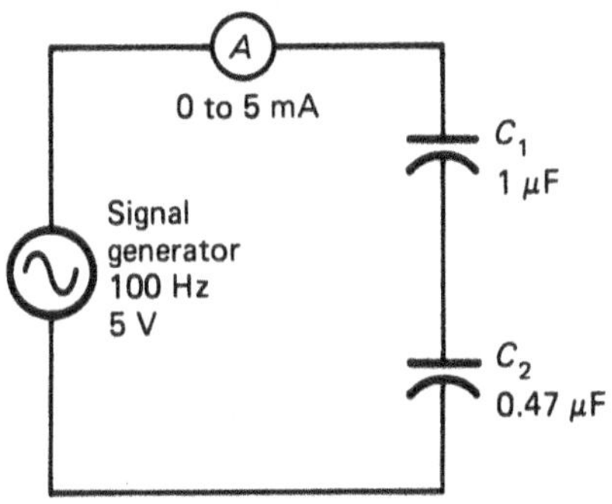

Fig. 10-12 Circuit for experiment 10-4, step 1.

- **a.** What is the equivalent capacitance of the circuit? (*Note:* Since the capacitors have 10 percent tolerances, consider C_2 to be 0.5 μF for all calculations.) ______________
- **b.** What is the total reactance of the circuit? ______________
- **c.** How much current should flow in the circuit? ______________
- **d.** Construct the circuit of Fig. 10-12 and measure the current. Does the measured current agree (within 10 percent) with the predicted current in step 1c? ______________
- **e.** How much voltage would you expect across C_1? ______________
- **f.** How much voltage would you expect across C_2? ______________
- **g.** To check your answers to steps 1e and 1f, measure and record these voltages. ______________ ______________
- **h.** In a series capacitor circuit, does the smallest or largest capacitor drop the most voltage? ______________
- **i.** Is the total reactance of the series circuit greater than or less than the reactance of either capacitor? ______________

2. Refer to the parallel circuit in Fig. 10-13.

- **a.** What is the equivalent (total) capacitance of the circuit? ______________
- **b.** What is the total reactance of the circuit? ______________

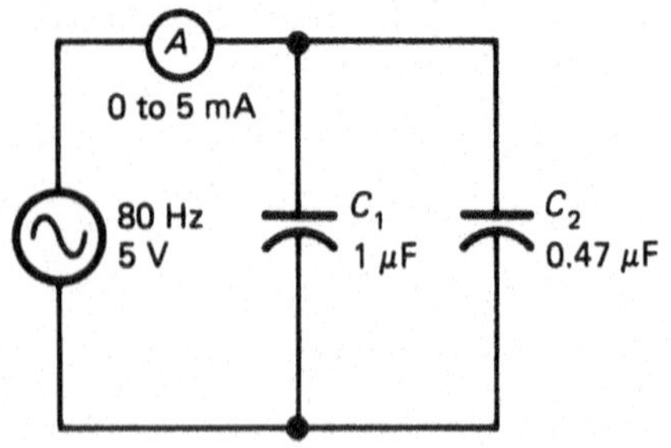

Fig. 10-13 Circuit for experiment 10-4, step 2.

c. What is the total current in the circuit? ______________

d. Construct the circuit shown in Fig. 10-13. Measure and record the total current. ______________

e. Do your answers to steps 2c and d agree within 10 percent? ______________

f. If they do, does it indicate that you used the correct procedures in answering steps 2a and b? ______________

g. Will C_1 or C_2 carry more current? ______________

h. Measure and record I_{C_1} and I_{C_2}.

______________ ______________

i. Does your answer to step 2h confirm your answer to step 2g? ______________

j. Do your answers to steps 2d and h confirm Kirchhoff's current law? ______________

3. As a review, answer the following questions:

a. Should two capacitors be connected in series or parallel to provide the most capacitance? ______________

b. Should two capacitors be connected in series or parallel to provide the most reactance? ______________

c. Should two capacitors be connected in series or parallel to draw the most current from the source? ______________

d. Does the larger or smaller of two parallel capacitors draw more current? ______________

NAME ____________________ DATE ____________

10-5 ADVANCED PROBLEMS

10-1. A series-parallel capacitor circuit is shown in Fig. 10-14. Determine the following for the circuit:

a. C_T (total capacitance across source) ____________

b. X_{C_T} ____________

c. I_{C_4} ____________

d. V_{C_3} ____________

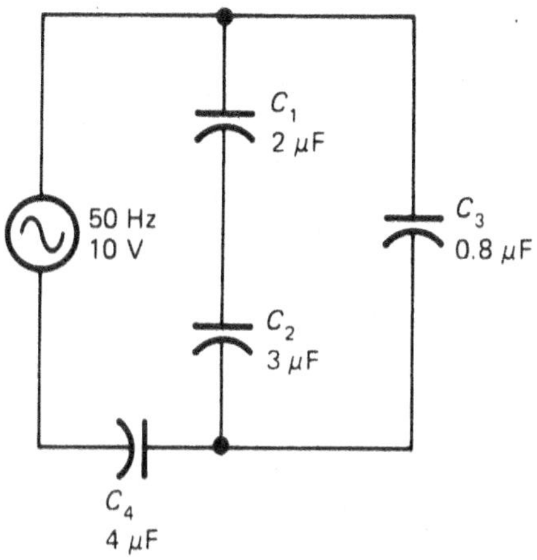

Fig. 10-14 Circuit for problem 10-1.

10-2. Refer to Fig. 10-15 and determine the following:

a. X_{C_1} ____________

b. X_{C_2} ____________

c. C_1 ____________

d. I_T ____________

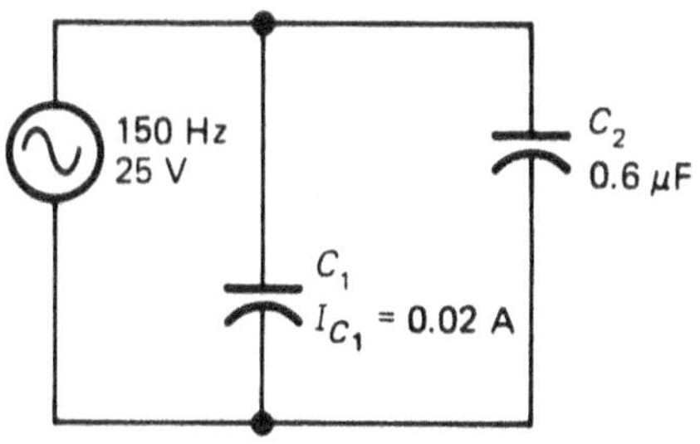

Fig. 10-15 Circuit for problem 10-2.

10-3. How much capacitance and charge are required to store 180 J of energy at 450 V? ____________ ____________

10-4. A capacitor is charged to 400 V by 5 C.

a. Determine the capacitance of the capacitor. ____________

b. How much energy is stored in the capacitor? ____________

10-6 MULTISIM: TROUBLESHOOTING 1

PURPOSE

This Multisim experiment provides experience in troubleshooting an *RC* circuit powered by a dc source (see Fig. 10-16). Assume the circuit components are soldered in place. Use only the *V* and *I* functions of the multimeter.

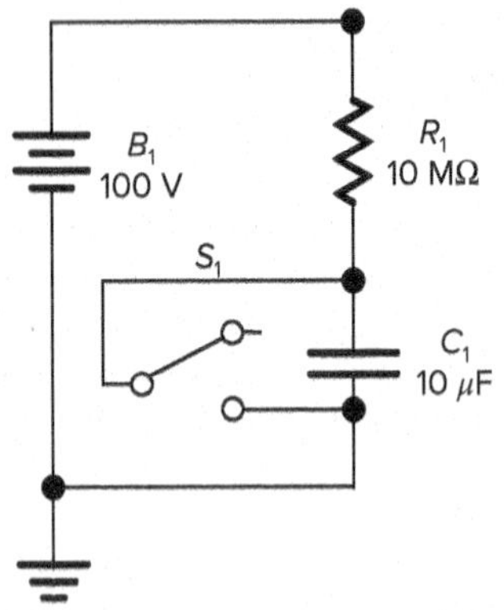

Fig. 10-16 Circuit for experiment 10-6.

MATERIALS

Qty.

Multisim
Multisim file for Fig. 10-16

PROCEDURE

1. **a.** Using Multisim, open the file for Fig. 10-16. S_1 is provided so that you can short out C_1 to discharge it when you want to remeasure charging time and/or voltage. (When S_1 is opened, C_1 starts to charge.) The multimeter available in Multisim has 10 MΩ of internal resistance on its dc voltage function.
   b. Calculate and record the time constant for the circuit. ______________
   c. Record the expected voltage across R_1 after one time constant. ______________
   d. When the multimeter (dc voltage function) is connected to R_1, what is the time constant? ______________
2. Activate the circuit, and determine which component or components are faulty and what is wrong with the component(s). Submit a report that details how you arrived at your conclusions.

NAME ______________________ DATE ______________

10-7 MULTISIM: TROUBLESHOOTING 2

PURPOSE

This Multisim experiment involves troubleshooting the circuit shown in Fig. 10-17. Assume the circuit components are soldered in place. Use only the *V* and *I* functions of the multimeter.

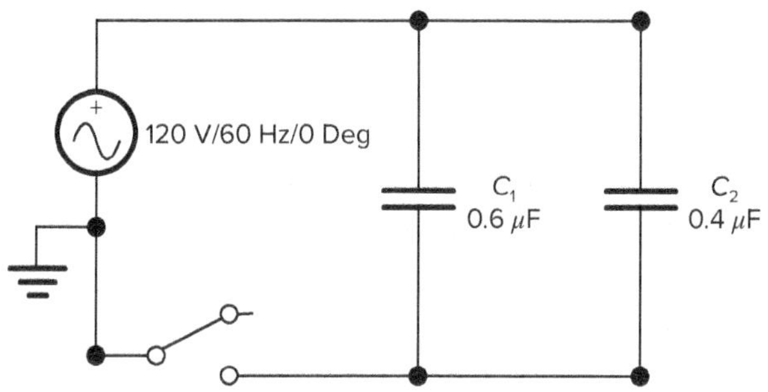

Fig. 10-17 Circuit for experiment 10-7.

MATERIALS

Qty.

Multisim
Multisim file for Fig. 10-17

PROCEDURE

1. Before making a measurement, you should know what value to expect from the measurement. For this circuit, what measurement(s) can you make, and what value(s) do you expect?

____

____

2. Using Multisim, open the file for Fig. 10-17. Activate the circuit, and determine which component or components are faulty and what is wrong with the component(s). Submit a report that details how you arrived at your conclusions.

10-8 MULTISIM: TROUBLESHOOTING 3

PURPOSE

This Multisim experiment involves troubleshooting the circuit shown in Fig. 10-18. Assume the circuit components are soldered in place. Use only the *V* and *I* functions of the multimeter.

MATERIALS

Qty.

Multisim
Multisim file for Fig. 10-18

PROCEDURE

1. Using Multisim, open the file for Fig. 10-18. Activate the circuit, and determine which component or components are faulty and what is wrong with the component(s). Submit a report that details how you arrived at your conclusions.

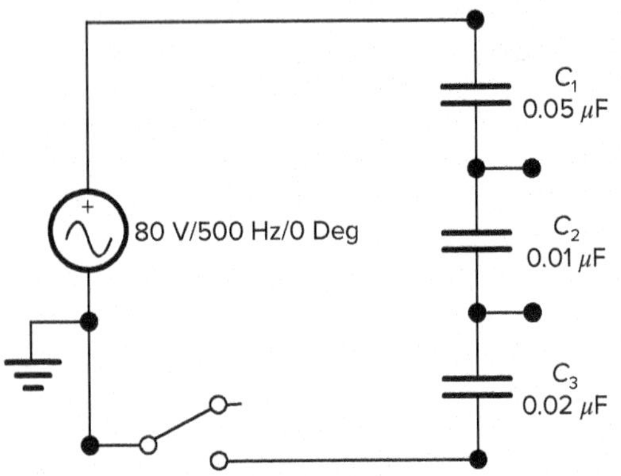

Fig. 10-18 Circuit for experiment 10-8.

NAME ____________________ DATE ____________

10-9 MULTISIM: DETERMINING CAPACITANCE WHEN FREQUENCY IS KNOWN

PURPOSE

Completing this experiment in Multisim will provide experience in using measured values to determine unknown values in a capacitor circuit.

MATERIALS

Qty.

Multisim
Multisim file for Fig. 10-19

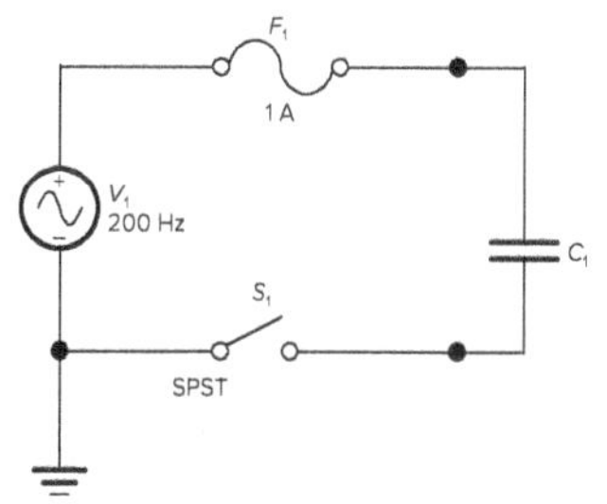

Fig. 10-19 Circuit for experiment 10-9.

PROCEDURE

1. Open the Multisim file for Fig. 10-19. You can measure the circuit current by placing the ammeter across the open switch.
2. Measure or calculate the following qualities for Fig. 10-19:

X_C = ______________

I_C = ______________

V_T = ______________

C = ______________

NAME ______________________________ DATE ______________

CHAPTER 11

Inductance

TEST: INDUCTANCE

For questions 1 to 23, determine whether each statement is true or false.

1. Ohmmeters cannot measure the reactance of an inductor. **1.** ______

2. The polarity of the cemf always opposes the action that created the cemf. **2.** ______

3. Inductance opposes a change in current when used in an ac circuit. **3.** ______

4. A straight length of conductor has both resistance and reactance. **4.** ______

5. An inductor has mutual inductance between its turns. **5.** ______

6. An inductor stores energy in its magnetic field. **6.** ______

7. Inductance converts electric energy into heat energy. **7.** ______

8. The cemf of an inductor opposes the source voltage when the current in the circuit is decreasing. **8.** ______

9. Adding turns to an inductor increases its inductance. **9.** ______

10. Increasing the space between the turns of an inductor increases its inductance. **10.** ______

11. Other factors being equal, a small-diameter coil will have less inductance than a large-diameter coil. **11.** ______

12. The inductance of a laminated-iron-core inductor decreases when its current increases. **12.** ______

13. The air-core symbol is used to represent a coil wound on a phenolic core. **13.** ______

14. Toroid cores are made from magnetic materials. **14.** ______

15. RF chokes usually have less than one base unit of inductance. **15.** ______

16. Mutual inductance can be reduced by magnetic shielding. **16.** ______

17. The dc resistance of an inductor can be measured with an ohmmeter. **17.** ______

18. The quality of an inductor depends on the frequency at which it is used. **18.** ______

19. Current decays more slowly in a resistor than it does in an inductor. **19.** ______

20. Inductive kick occurs when an inductive circuit is closed. **20.** ______

21. It is impossible for the cemf of an inductor to exceed the source voltage under any conditions. **21.** ______

22. At any frequency, a small inductance produces more reactance than a large inductance. **22.** ______

23. Current induced in the iron core of an inductor causes the inductance to increase. **23.** ______

For questions 24 to 30, supply the missing word or phrase in each statement.

24. The resistance of a coil is greater than its ohmic resistance because of the ______. **24.** ______

25. The induced voltage in an inductor is called ______. **25.** ______

26. Movable slug-type cores are used in ______ inductors. **26.** ______

27. Filter chokes for power-frequency circuits usually use ______________ cores. 27. ______________

28. ______________-core inductors usually have a voltage rating as well as an inductance rating. 28. ______________

29. Inductive current ______________ inductive voltage by ______________ degrees. 29. ______________ ______________

30. Litz wire is used to reduce the ______________. 30. ______________

For questions 31 to 42, choose the letter that best completes each statement.

31. The symbol for inductance is 31. ______________
   a. *H*
   b. *L*
   c. *I*
   d. *Q*

32. An inductor should not be called a 32. ______________
   a. Choke
   b. Coil
   c. Reactor
   d. Turn

33. The opposition of inductance to alternating current is called 33. ______________
   a. Reactance
   b. Reluctance
   c. Permeability
   d. Impedance

34. Inductance is that electrical property that 34. ______________
   a. Converts electric energy into heat energy
   b. Converts heat energy into electric energy
   c. Opposes changes in current
   d. Opposes changes in resistance

35. Flux from one turn of an inductor cutting the adjacent turns of the same inductor is an example of 35. ______________
   a. Impedance
   b. Reactance
   c. Self-inductance
   d. Mutual inductance

36. The symbol for reactance is 36. ______________
   a. *X*
   b. *H*
   c. *L*
   d. *Q*

37. The base unit of inductance is the 37. ______________
   a. Henry
   b. Ohm
   c. Tesla
   d. Lenz

38. The base unit of reactance is the 38. ______________
   a. Henry
   b. Ohm
   c. Tesla
   d. Lenz

39. An inductor is wound on a paper tube. Its inductance is greatest when the tube contains 39. ______________
   a. Air
   b. Brass
   c. Phenolic
   d. Ferrite

40. An inductor is wound on a paper tube. Its inductance is least when the tube contains 40. ______________
   a. Air
   b. Brass
   c. Silicon steel
   d. Ferrite

NAME ____________________ DATE ____________

41. An inductor used to help smooth a pulsating direct current is **41.** ________
   a. An RF coil
   b. A filter choke
   c. An RF choke
   d. A molded choke

42. Iron-core inductors are not rated for **42.** ________
   a. Power
   b. Quality
   c. Current
   d. Voltage

For questions 43 to 54, solve each problem. Be sure to include units (watts, degrees, etc.) in your answer when appropriate.

43. What is the reactance of an inductor that draws 0.25 A from a 60-V, 200-Hz source? **43.** ________

44. What is the reactance of a 2.5-mH inductor connected to a 100-kHz, 20-V source? **44.** ________

45. What is the quality of a 0.01-H inductor that has a resistance of 40 Ω at 50 kHz? **45.** ________

46. What is the equivalent inductance of a 4-H inductor and a 6-H inductor connected in parallel? **46.** ________

47. Suppose the parallel inductors in question 46 were connected to an ac source. Which inductor would draw more current? **47.** ________

48. Refer to Fig. 11-1. Which inductor will drop the most voltage? **48.** ________

49. Refer to Fig. 11-1. What is the total inductance of the circuit? **49.** ________

50. Refer to Fig. 11-1. What is the current in the circuit? **50.** ________

51. Determine the frequency that is needed to produce a reactance of 2500 Ω with a 3.3-mH inductor. **51.** ________

52. Determine the resistance of a 0.05-H inductor that has a Q of 75 at 40 kHz. **52.** ________

53. Determine the total current, total reactance, and total inductance for the values given in Fig. 11-1 when the inductors are connected in parallel rather than in series.

53. I_T = ________
X_{L1} = ________
L_T = ________

54. Determine L_T, X_{L_T}, I_T, and $I_{0.4\,H}$ (the current through the 0.4-H inductor) in Fig. 11-1 when the 0.4-H inductor is paralleled with a 0.6-H inductor.

54. L_T = ________
X_{L_T} = ________
I_T = ________
$I_{0.4\,H}$ = ________

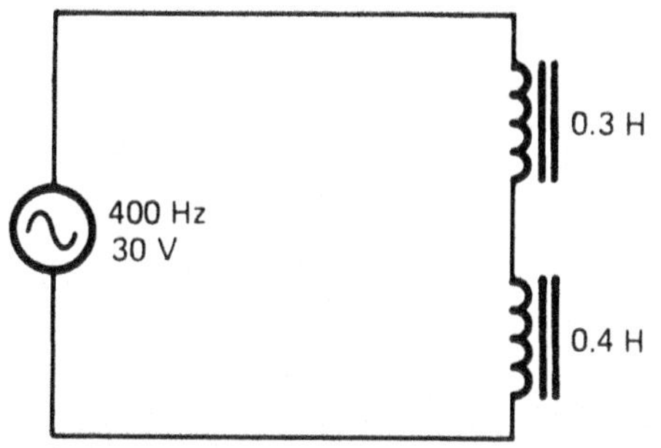

Fig. 11-1 Circuit for questions 48 through 50, 53, and 54.

NAME ______ DATE ______

11-1 LAB EXPERIMENT: INDUCTORS IN AC AND DC CIRCUITS

PURPOSE

This experiment will visually demonstrate the major characteristics of inductors in both ac and dc circuits.

MATERIALS

Qty.		Qty.	
1	choke, 0.5 H, 300 mA dc, 30-Ω dc resistance	1	lamp, no. 47, 6.3 V, 0.15 A
1	choke, 2.5 mH, 125 mA, 38-Ω dc resistance	1	lamp holder for no. 47 lamp
1	resistor, 22 Ω, 10%, 1 W	1	power supply, 6 to 7 V dc
		1	power supply, 6 to 7 V ac
		1	VOM or DMM

PROCEDURE

1. Construct the circuit shown in Fig. 11-2(*a*). Note that the lamp is not at full brightness. This is because of the dc, or ohmic, resistance of the inductor. Remove the inductor from the circuit.

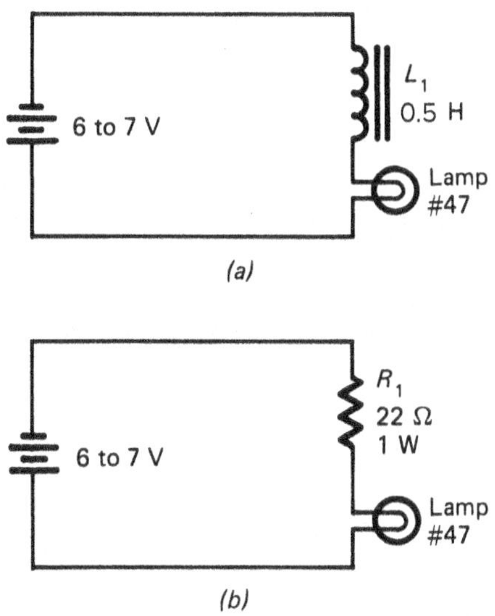

Fig. 11-2 Circuits for experiment 11-1, steps 1 and 3.

a. Measure and record the (ohmic) resistance between the inductor's terminals. ______

b. Now construct the circuit in Fig. 11-2(*b*). This replaces the inductor with a 22-Ω, 1-W resistor. Does the lamp have about the same brightness with the resistor in the circuit as it did with the inductor in the circuit? ______

c. Do the results above indicate that the only opposition an inductor has to pure direct current is its ohmic (dc) resistance? ______

2. Construct the circuit of Fig. 11-3(*a*) by changing from a dc to an ac supply.

a. Does the lamp have the same brightness as it did with the dc supply? ______

b. Does the result in step 2a indicate that a resistor behaves the same in dc circuits as it does in ac circuits? ______

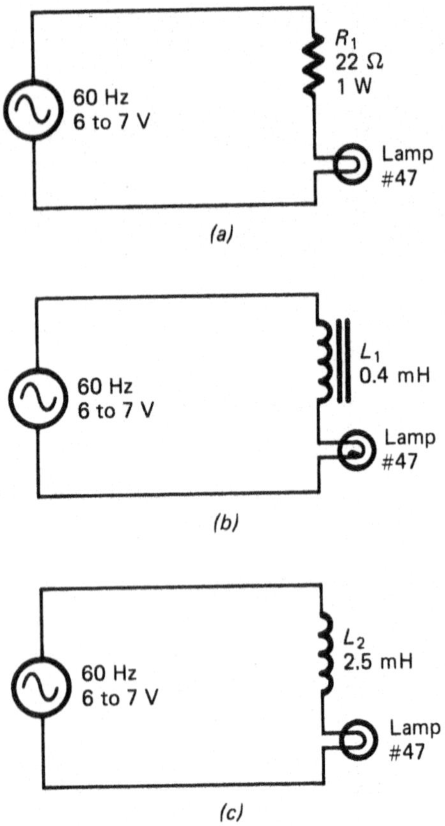

Fig. 11-3 Circuits for experiment 11-1, steps 2 and 4.

c. Now replace the resistor with the 0.5-H inductor to provide the circuit shown in Fig. 11-3(*b*). Does the lamp now light? ____________ Why?

____

____

d. Do the results in steps 2a and c indicate that an inductor offers more opposition to alternating current than just its ohmic resistance?

____

e. What is this additional opposition called? ____________

f. Is there any current flowing through the lamp in Fig. 11-3(*b*)? ____________ Check your answer by measuring the ac voltage across the lamp. (If current is flowing, there must be a voltage drop.)

g. Would a smaller inductance allow more current to flow through the lamp? ____________

h. Replace the 0.5-H inductor with a 2.5-mH inductor, as in Fig. 11-3(*c*). Does the lamp light now? ____________ Why?

____

____

i. Does a 0.5-H inductor or a 2.5-mH inductor have the greater reactance?

____

3. This section is concerned with observing inductive kick. Reconnect the circuit of Fig. 11-2(*b*). Quickly make and break (close and open) the circuit by interrupting the connection at one end of the resistor. Note the size of the spark that occurs as the circuit is interrupted. Now, reconnect the circuit of Fig. 11-2(*a*), and observe the spark when this circuit is interrupted. Which circuit provided the larger spark? ____________ Why?

____

____

4. As a review, answer the following questions:
   a. Does an ohmmeter measure the reactance of an inductor?

   ______

   b. How much reactance does the inductor in Fig. 11-3(*b*) have?

   ______

   c. How much reactance does the inductor in Fig. 11-3(*c*) have?

   ______

   d. What type of core does L_1 in Fig. 11-3(*b*) have? ______

11-2 LAB EXPERIMENT: INDUCTIVE KICK

PURPOSE

This experiment will show visual evidence of inductive kick. It will provide proof that a cemf can far exceed the source voltage that created it when the circuit is opened. As an option, this experiment may be more suitable as a classroom or laboratory demonstration led by an instructor.

MATERIALS

Qty.		Qty.	
1	choke, 0.5 H, 300 mA dc, 30-Ω dc resistance	1	dry cell, 1.5 V, size D
3	neon lamps, NE-2	1	switch, SPST, 1 A, 125 V

INTRODUCTION

NE-2 lamps will be used to indicate the magnitude of the inductive kick (cemf) of an inductor. It requires about 80 V (see Fig. 11-4) to fire (ionize) an NE-2. Once the lamp fires, it requires about 70 V to maintain ionization. The current through the lamp, once it fires, rises very high unless it is limited by the source or by a series resistor. In this activity the source will be the energy stored in the inductor's magnetic field. This source can provide only a small current at the 70 V required to keep the NE-2 ionized.

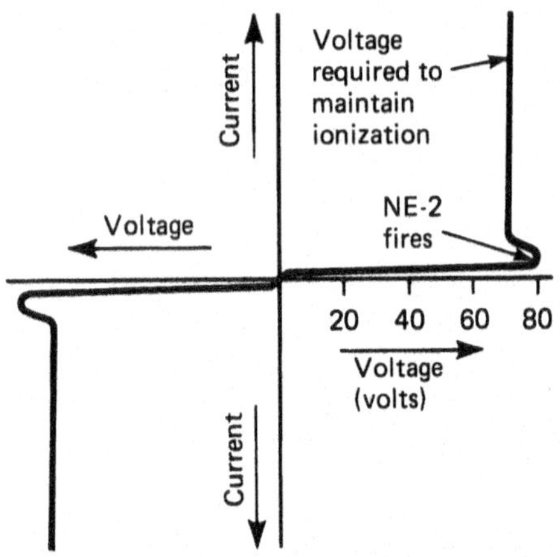

Fig. 11-4 Characteristic curve of an NE-2 neon lamp.

Throughout this experiment, S_1 in Fig. 11-5 should be closed only when called for in the procedure. Leaving S_1 closed for extended periods will only shorten the useful life of the dry cell.

PROCEDURE

1. Construct the circuit shown in Fig. 11-5(*a*); have S_1 open. While observing the NE-2 lamp, close S_1 for a moment and then open S_1.
   a. Did the NE-2 fire when S_1 first closed? ______________
   b. Did the NE-2 fire while S_1 remained closed? ______________
   c. How much voltage must have been across the NE-2 while S_1 was closed? ______________
   d. Did the lamp ionize when the switch was opened? ______________
   e. How much voltage appeared across the inductor and the NE-2 when the switch was open? (Fig. 11-4) ______________
   f. What caused this relatively high voltage to appear across the NE-2? ______________________________

NAME ______________________________ DATE ______________

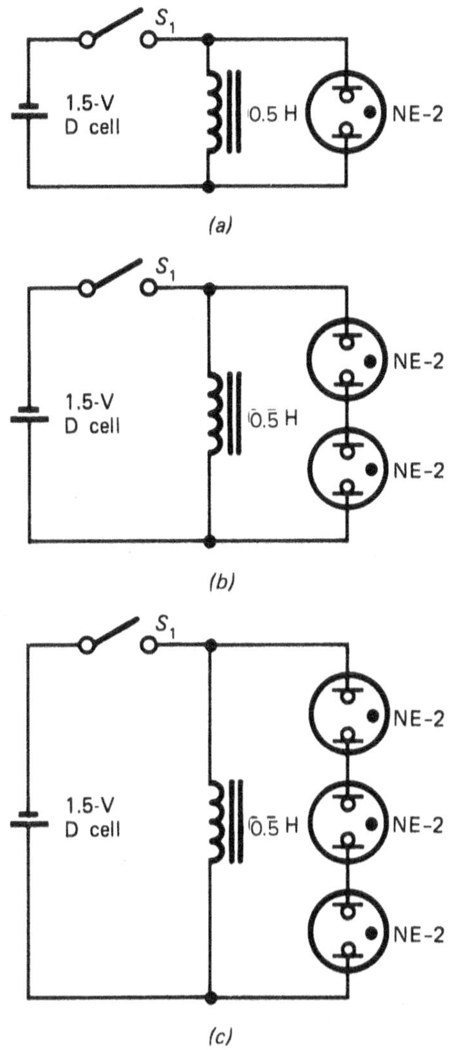

Fig. 11-5 Circuits for experiment 11-2.

2. The NE-2 in Fig. 11-5(*a*) uses up the inductor's stored energy once it fires. It also limits the inductor's cemf to the 80 V required for ionization. Do you think that the inductor's cemf can rise high enough to fire two NE-2 lamps connected in series? (This will require about 160 V.) Construct the circuit of Fig. 11-5(*b*) and see if it will. Does the inductive kick fire both lamps? ______________

3. **a.** How much voltage will be required to fire the three NE-2 lamps in Fig. 11-5(*c*)? ______________

b. Construct the circuit and see if the cemf will fire the three lamps. Did it? ______________

c. The main source voltage for this circuit is the 1.5-V cell. How many times larger is the inductor's cemf than the source voltage when S_1 is opened?

____

NAME ______________________ DATE ______________

11-3 LAB EXPERIMENT: RELATIONSHIP BETWEEN f AND X_L

PURPOSE

This experiment will demonstrate the effect of frequency on inductive reactance. It will also provide experience in calculating reactance by the reactance formula and by Ohm's law.

MATERIALS

Qty.		Qty.	
1	choke, 2.5 mH, 125 mA, 38-Ω dc resistance	1	signal generator, 5 V, 25- to 100-kHz sinusoidal output
1	VOM or DMM	1	resistor, 100 Ω, 5%, ½ W

INTRODUCTION

Ammeters that will measure 25-kHz currents are not common. The circuit in this experiment operates at 25 kHz and above. Therefore, we are going to indirectly determine the circuit current. That is, we will measure the voltage across a known resistance and then use Ohm's law to calculate the current ($I_R = V_R/R$). The circuit (Fig. 11-6) used in this experiment is a series circuit. Thus, the current through the resistor is also the current through the inductor ($I_R = I_L$).

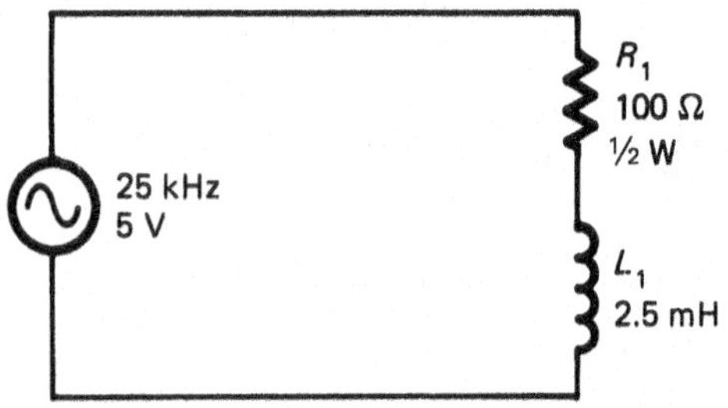

Fig. 11-6 Circuit for experiment 11-3.

The inductive reactance X_L can now be experimentally determined by using the measured voltage across the inductor and Ohm's law ($X_L = V_L/I_L$).

You may notice in this experiment that the arithmetic sum of V_R and V_L is greater than the supply voltage. Do not be concerned; the arithmetic sum should be greater. Resistive voltage and inductive voltage must be treated as phasors. Phasors are covered in more detail in Chap. 13.

PROCEDURE

1. Construct the circuit shown in Fig. 11-6. Be sure the generator is set for 25 kHz and the output voltage is 5 V. Measure the output voltage with the DMM. Measure the voltage across R_1 and record it in the data table of Fig. 11-7. Now calculate the current through the resistor and enter it in the data table. Next measure and record the voltage across the inductor. Finally, calculate the reactance and enter it in the table (remember $I_R = I_L$).

   Check to see if the reactance determined above agrees with the value calculated by the reactance formula. Calculate and record the reactance using the formula $X_L = 6.28fL$. ______________

   The reactance calculated here should be within 15 percent of the reactance you listed in the data table. If it is not, recheck your measurements and your calculations.

Frequency f	Voltage across R_1 V_{R_1} (measured)	Current through R_1 I_{R_1}	Voltage across L_1 V_{L_1} (measured)	Inductive reactance X_{L_1}
25 kHz				
50 kHz				
100 kHz				

Fig. 11-7 Data table for experiment 11-3.

2. Change the frequency of the generator to 50 kHz. Measure the output voltage of the generator. Readjust the output control (if necessary) to obtain exactly 5 V. You have doubled the frequency of the generator and held the voltage constant.

a. What do you predict will happen to the reactance?

____

____

b. What do you predict will happen to the current?

____

____

c. Make the measurements and calculations necessary to complete the second row of the data table. Does your data table support your predictions in steps 2a and b? ____________

3. Change the generator frequency to 100 kHz. Readjust the output for 5 V if necessary. From the reactance you determined in the first two rows of the data table, predict and record the reactance you will have at 100 kHz. ____________ Complete the third row of the data table of Fig. 11-7 to check your prediction.

11-4 LAB EXPERIMENT: SERIES AND PARALLEL INDUCTORS

PURPOSE

This experiment will illustrate the current and voltage distribution in series and parallel inductance circuits. It will also verify the series and parallel formulas used for determining total inductance and total reactance.

MATERIALS

Qty.		Qty.	
2	chokes, 2.5 mH, 125 mA, 38-Ω dc resistance	1	signal generator, 5 V, 50-kHz sinusoidal output
1	VOM or DMM	1	resistor, 51 Ω, 5%, ½ W
		2	resistors, 100 Ω, 5%, ½ W

INTRODUCTION

The circuits in this experiment are all operated at 50 kHz. Ammeters that measure current at 50 kHz are not common. Therefore, the current in these circuits will be indirectly determined by using Ohm's law. A small resistor is put in the circuit in which we wish to measure current. Then the voltage across the resistor is measured, and Ohm's law ($I_R = V_R/R$) is used to calculate the current.

Once we have determined the current in a circuit, we can again use Ohm's law to determine the inductive reactance. The formula is

$$X_L = \frac{V_L}{I_L}$$

Of course, I_L will be the same as I_R when the resistor is in series with the inductor.

You may notice that the arithmetic sum of V_R and V_L is greater than the source voltage. Resistive voltage and inductive voltage must be treated as phasors. Phasors are covered in Chap. 13.

PROCEDURE

1. **a.** The reactance of a 2.5-mH inductor at 50 kHz is 785.4 Ω. How much reactance would you expect from two 2.5-mH inductors connected in parallel? ______________

   b. Check your response by constructing the circuit in Fig. 11-8(*a*) and determining its total reactance. First measure and record the voltage across the resistor. ______________

   c. Then calculate and record the current through the resistor ($I_{R_1} = V_{R_1}/R_1$). This is the total current. ______________

   d. The total current splits at the junction of L_1 and L_2 and goes through the parallel inductors. Therefore, the total inductive reactance can be found

$$X_{L_T} = \frac{V_L}{I_T}$$

   Measure and record the voltage across either inductor. ______________

   e. Calculate and record the total reactance. Is your answer within 15 percent of your answer to step 1a? ______________ If not, check your work!

   f. Calculate and record the total inductance of L_1 and L_2 in parallel. ______________

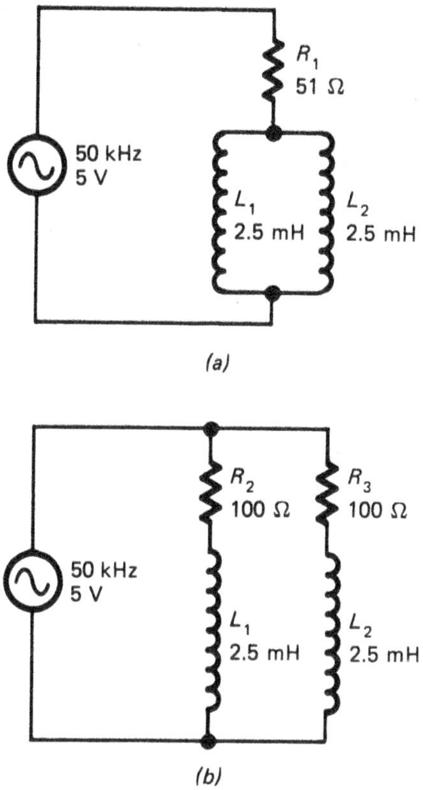

Fig. 11-8 Circuits for experiment 11-4, steps 1 and 2.

$$L_T = \frac{L_1 \times L_2}{L_1 + L_2}$$

g. Now calculate and record the inductive reactance of this total inductance by using the reactance formula. This value of X_{L_T} should be the same (within 15 percent) as you found in step 1e. _______________

2. a. To check the current distribution in a parallel inductor circuit, set up the circuit shown in Fig. 11-8(*b*). Measure and record the voltage across R_2 and then across R_3. _______________ _______________

b. Since $R_2 = R_3$, do your results show that $I_{L_1} = I_{L_2}$ (within component tolerances, ±15 percent)? _______________

c. If L_1 were a 5-mH inductor, would $I_{L_1} = I_{L_2}$? _______________

3. a. Refer to the circuit in Fig. 11-9. Would you expect V_{L_1} to be equal to V_{L_2}? _______________ Why?

____

____

b. What should the total reactance of the circuit be? _______________

c. Construct the circuit in Fig. 11-9 and check your answers to steps 3a and b. Measure and record V_{L_1}. _______________

d. Measure and record V_{L_2}. _______________

e. Measure and record V_{R_1}. _______________

f. Calculate and record I_T ($I_T = I_{R_1} = V_{R_1}/R_1$). _______________

g. Calculate and record X_{L_T}. _______________ Note:

$$X_{L_T} = \frac{V_{L_T}}{I_T} \text{ and } V_{L_T} = V_{L_1} + V_{L_2}$$

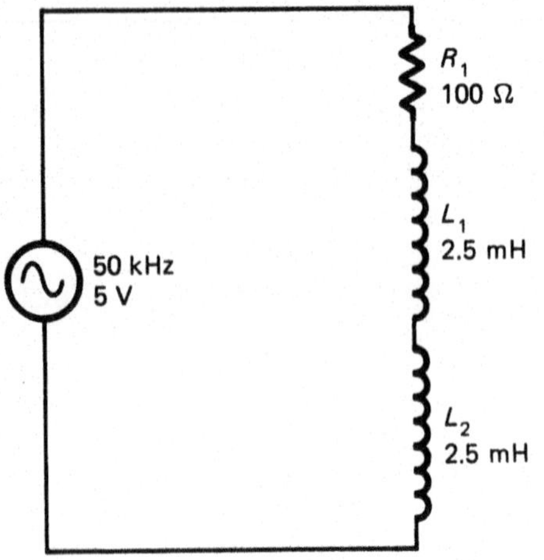

Fig. 11-9 Circuit for experiment 11-4, step 3.

h. Do your results show that both inductance and inductive reactance are additive when in series? ______________

NAME ____________________ DATE ____________

11-5 ADVANCED PROBLEMS

11-1. Refer to the circuit in Fig. 11-10(*a*), and calculate and record the following:

a. L_T ____________ **d.** I_{L_1} ____________
b. X_{L_T} ____________ **e.** I_{L_2} ____________
c. X_{L_1} ____________

11-2. Refer to the circuit in Fig. 11-10(*b*), and calculate and record the following:

a. X_{L_2} ____________ **c.** V_{L_1} ____________
b. X_{L_T} ____________ **d.** L_1 ____________

11-3. Refer to the circuit in Fig. 11-10(*c*), and calculate and record the following:

a. X_{L_1} ____________ **d.** X_{L_3} ____________
b. X_{L_2} ____________ **e.** L_3 ____________
c. I_{L_2} ____________

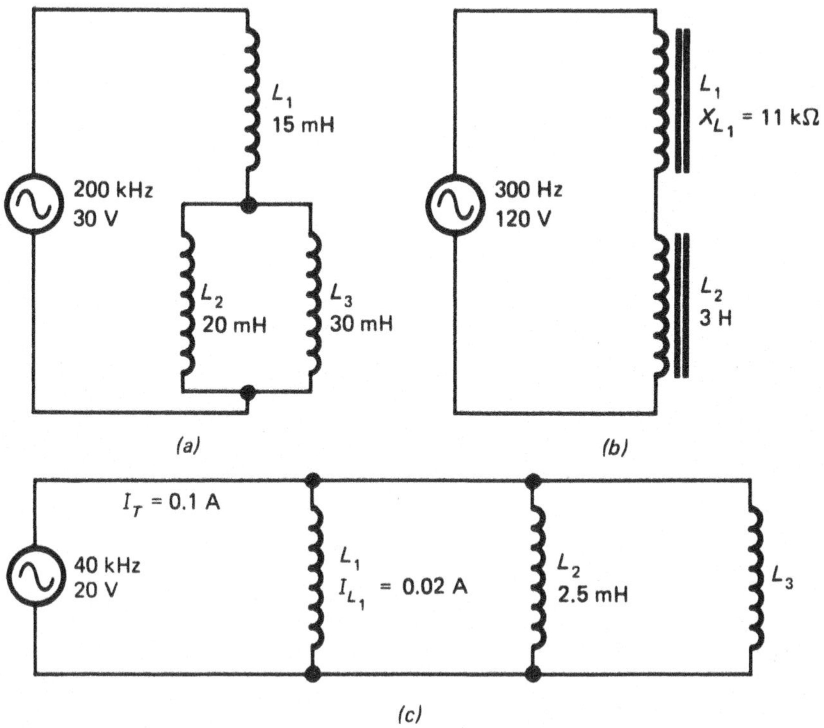

Fig. 11-10 Circuits.

11-6 MULTISIM: *L*, *f*, AND X_L RELATIONSHIPS

PURPOSE

This experiment is a repeat of experiment 11-3 except that Multisim will be used instead of actual electronic components.

MATERIALS

Qty.

Multisim

PROCEDURE

1. Using Multisim, construct the circuit shown in Fig. 11-11. Activate the circuit.
   a. Measure and record the voltage across and the current through the inductor. ______________ ______________
   b. Using the above measured values, calculate and record X_L. ______________
   c. Using the reactance formula, calculate and record X_L. ______________
   d. If the values of X_L determined in steps 1b and c do not agree within ±2 percent, check your measurements and calculation.
2. Predict (and record your prediction) what will happen to the current and reactance when you change the frequency of the generator to 50 kHz.
   I ____
   X_L ____
3. Change the frequency to 50 kHz and check predictions.
   a. Measure and record the voltage. ______________
   b. Measure and record the current. ______________
   c. Using the measured value, calculate and record the reactance. ______________
4. What value of current would you expect if L_1 were changed to 5 mH and *f* were changed back to 25 kHz? ______________
5. Make the changes specified in step 4 and measure and record the current. ______________

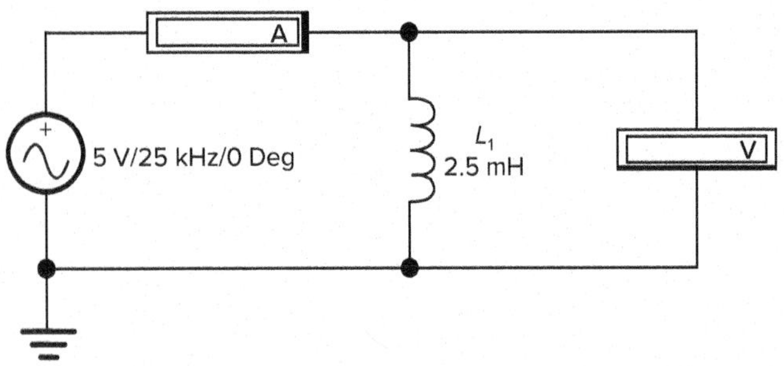

Fig. 11-11 Circuit for experiment 11-6.

NAME ______________________ DATE ______________

11-7 MULTISIM: *L* IN SERIES AND PARALLEL

PURPOSE

This experiment is a repeat of experiment 11-4 except that Multisim will be used instead of actual electronic components.

MATERIALS

Qty.

Multisim

PROCEDURE

1. **a.** Calculate and record the reactance of a 2.5-mH inductor at 50 kHz. ______________

   b. From the above calculation, predict the total reactance of the circuit in Fig. 11-12. ______________

   c. Calculate and record the total inductance for the circuit in Fig. 11-12. ______________

   d. Using your answer to step 1c, calculate and record the total reactance for the circuit in Fig. 11-12. ______________

   e. Do the values of steps 1b and d agree? ______________ (If not, rethink your answers!)

   f. Calculate and record the total current for the circuit in Fig. 11-12. ______________

   g. Should $I_{L_1} = I_{L_2}$ in the circuit in Fig. 11-12? ______________

   h. If L_1 were a 5-mH inductor would $I_{L_1} = I_{L_2}$? ______________

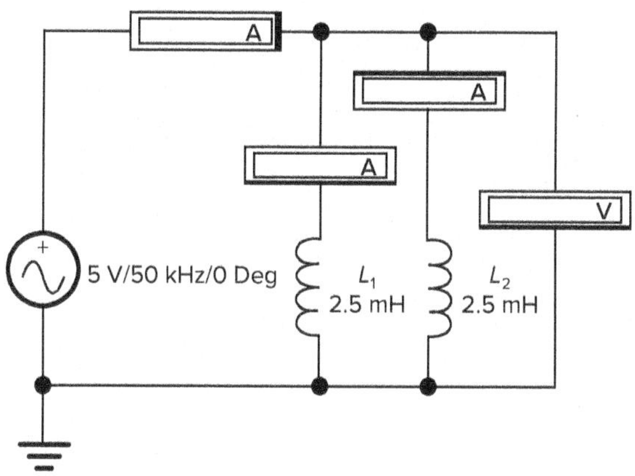

Fig. 11-12 Multisim circuit for experiment 11-7, steps 1 and 2.

2. Using a circuit simulation program, construct and activate the circuit shown in Fig. 11-12.

   a. Record the measured total voltage. ______________

   b. Record the measured total current. ______________

   c. Using measured values, calculate and record the total reactance. ______________ (Check your answer against that given in step 1d.)

   d. Do the measured values of I_{L_1} and I_{L_2} support your answer given in step 1g? ______________

   e. What value of I_T would you expect if L_1 were changed to 5 mH? ______________

   f. Change L_1 to a 5-mH inductor and then measure and record I_T. ______________

3. **a.** For the circuit shown in Fig. 11-13, calculate and record values for the following:

   X_{L_1} ______________ I_T ______________

   X_{L_T} ______________ V_{L_1} ______________

   b. Should $V_{L_1} = V_{L_2}$ in this circuit? ______________

4. Using Multisim, construct and activate the circuit shown in Fig. 11-13.

   a. Record the following measured values:

   V_T ______________ V_{L_1} ______________

   I_T ______________ V_{L_2} ______________

   b. Using measured values, calculate and record the following quantities:

   X_{L_T} ______________ X_{L_1} ______________

   c. Do the values you measured and calculated in steps 4a and b agree (within 2 percent) with your answers in steps 3a and b? ______________ (If not, there is a problem!)

   d. If L_2 in Fig. 11-13 were changed to 10 mH, predict the following:

   V_{L_1} ______________ I_T ______________

   e. Change L_2 to 10 mH. Do the measured values support your answers to step d above? ______________

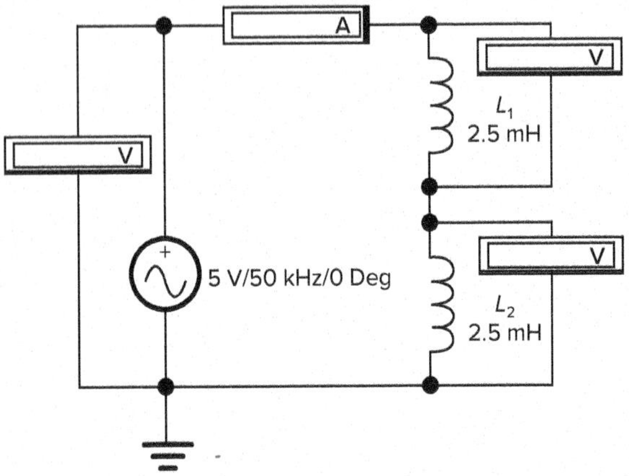

Fig. 11-13 Multisim circuit for experiment 11-7, steps 3 and 4.

NAME ______________________ DATE ______________

11-8 LAB EXPERIMENT: TROUBLESHOOTING 1

PURPOSE

Troubleshooting this inductor circuit will give you more experience and practice in both measuring quantities and predicting and/or calculating expected values. Predicting and measuring values are the essence of troubleshooting.

Qty.

Multisim
Multisim file for Fig. 11-14.

PROCEDURE

1. For the circuit shown in Fig. 11-14, calculate and record the following:
   X_{L_1} ______________ X_{L_2} ______________
   I_T ______________ V_{L_1} ______________
2. Using Multisim, open the file for Fig. 11-14 and activate the circuit. Determine which component or components are faulty and what is wrong with the component(s). Submit a report detailing how you found the faulty component(s).

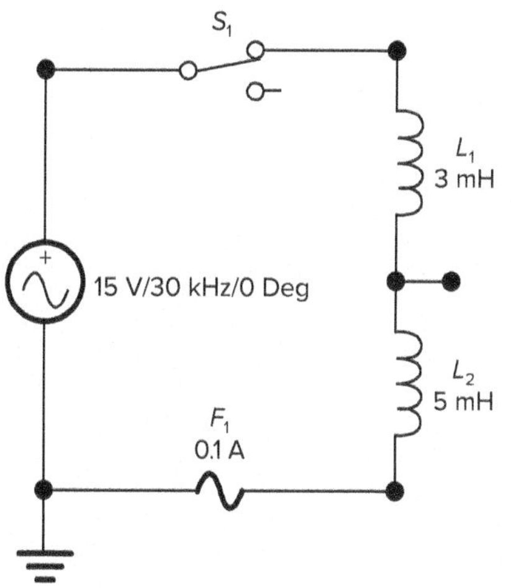

Fig. 11-14 Circuit for experiment 11-8.

11-9 LAB EXPERIMENT: TROUBLESHOOTING 2

PURPOSE

Troubleshooting this inductor circuit will give you more experience and practice in both measuring quantities and predicting and/or calculating expected values. Predicting and measuring values are the essence of troubleshooting.

Qty.

Multisim
Multisim file for Fig. 11-15

PROCEDURE

1. For the circuit shown in Fig. 11-15, calculate and record the following:

   X_{L_1} ______________ I_T ______________
   X_{L_2} ______________ V_{L_1} ______________
   X_{L_3} ______________ I_{L_2} ______________

2. Using Multisim, open the file for Fig. 11-15 and activate the circuit. Determine which component or components are faulty and what is wrong with the component(s). Submit a report detailing how you found the faulty component(s).

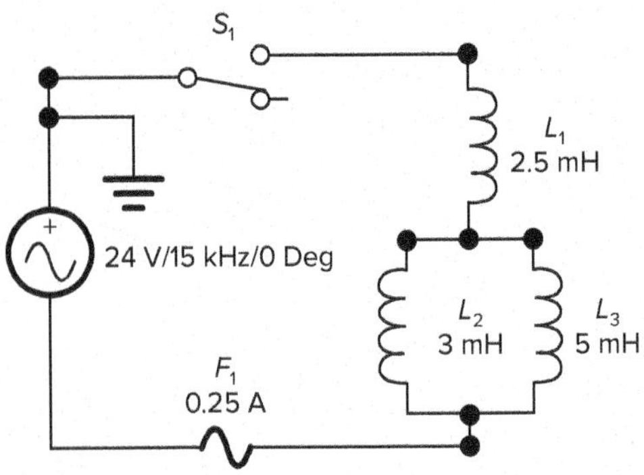

Fig. 11-15 Circuit for experiment 11-9.

NAME ______________________________ DATE ______________

11-10 MULTISIM: DETERMINING INDUCTANCE WHEN FREQUENCY IS KNOWN

PURPOSE

This experiment provides experience in using measured values to determine inductance and inductive reactance.

Qty.

Multisim
Multisim file for Fig. 11-16

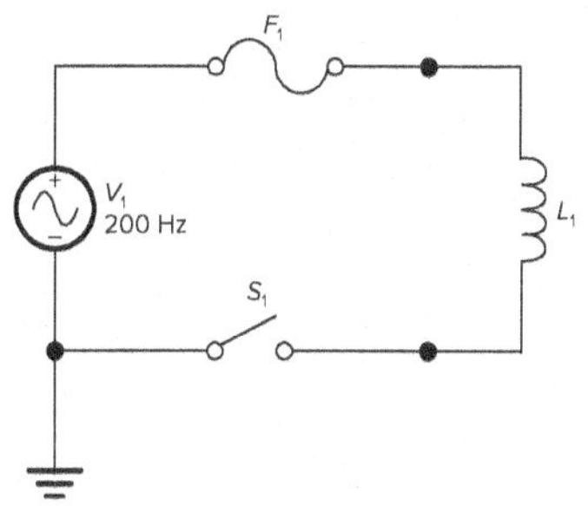

Fig. 11-16 Circuit for experiment 11-10.

PROCEDURE

1. Open the Multisim file for Fig. 11-16, and measure the current by placing the ammeter across the open switch.
2. Measure or calculate the following for Fig. 11-16:

   X_L = ______________

   I_L = ______________

   V_T = ______________

   L = ______________

NAME ______________________________ DATE ______________

CHAPTER 12

Transformers

TEST: TRANSFORMERS

For questions 1 to 27, determine whether each statement is true or false.

1. The same base unit is used for both mutual inductance and self-inductance. **1.** ________

2. Transformers sometimes have a nonmagnetic shield around the coil. **2.** ________

3. The primary of a transformer converts electric energy into magnetic energy. **3.** ________

4. The designated primary of a transformer can be used as a secondary. **4.** ________

5. Air-core transformers have a higher coefficient of coupling than laminated-iron-core transformers do. **5.** ________

6. Leakage flux reduces the coefficient of coupling. **6.** ________

7. The voltage ratio of a transformer must be greater than the turns ratio. **7.** ________

8. The turns-per-volt ratio of the secondary is usually smaller than the turns-per-volt ratio of the primary. **8.** ________

9. On a multisecondary transformer, each secondary has the same turns-per-volt ratio. **9.** ________

10. A 400-W transformer has fewer turns per volt than a 30-W transformer. **10.** ________

11. A coefficient of coupling that is less than 1 means that some of the primary flux does not cut the secondary coil. **11.** ________

12. Both the core and the coils in a magnetic-core transformer cause the transformer to heat up when operating. **12.** ________

13. A core material with a narrow hysteresis loop should be used to reduce transformer losses. **13.** ________

14. Eddy currents are induced in the secondary of a transformer. **14.** ________

15. The I^2R loss in a transformer is also called copper loss. **15.** ________

16. The primary and secondary currents in a transformer create magnetizing forces that aid each other. **16.** ________

17. Autotransformers provide electrical isolation between the primary and the secondary. **17.** ________

18. From the primary side of a step-down transformer, the secondary load appears to be greater than its actual value. **18.** ________

19. Impedance matching provides maximum efficiency as power is transferred from source to load. **19.** ________

20. The turns ratio and the impedance ratio of a transformer are numerically equal. **20.** ________

21. If the current ratings of a transformer are not exceeded, other transformer ratings will not be exceeded. **21.** ________

22. Rated secondary voltage is the voltage provided by the secondary before it is loaded. **22.** ________

23. Operating a transformer with less than rated voltage on the primary doesn't harm the transformer. **23.** ________

24. Dual primary windings are connected in parallel for operation on the lowest-rated primary voltage. 24. ______________

25. If a primary is wye-connected, the secondary must be delta-connected. 25. ______________

26. The transformer in Fig. 12-1(*a*) has a wye primary. 26. ______________

27. The transformer in Fig. 12-1(*b*) is a step-down autotransformer. 27. ______________

For questions 28 to 33, supply the missing word or phrase in each statement.

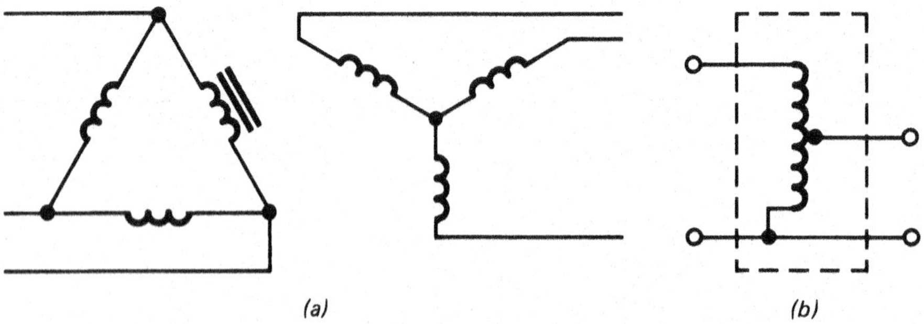

Fig. 12-1 Symbols for questions 26 and 27.

28. The primary and secondary of a transformer are coupled together by ______________. 28. ______________

29. For a given primary voltage, the secondary voltage will increase if the coefficient of coupling is ______________. 29. ______________

30. A(n) ______________ transformer has less secondary voltage than it has primary voltage. 30. ______________

31. Before going to the power lines, the voltage from the power plant generator is ______________. 31. ______________

32. If a transformer steps up voltage, it will ______________ current. 32. ______________

33. Hysteresis is caused by the ______________ in the magnetic core material. 33. ______________

For questions 34 to 47, choose the letter that best completes each statement.

34. Transformers operate on the principle of 34. ______________
    a. Reactance
    b. Reluctance
    c. Mutual inductance
    d. Self-inductance

35. A transformer always 35. ______________
    a. Changes voltage levels
    b. Matches impedances
    c. Electrically isolates two circuits
    d. Converts some electric energy to heat energy

36. Power loss in the core of a transformer is caused by 36. ______________
    a. Reactance of the coils
    b. Reluctance of the core
    c. Permeability of the core
    d. Eddy currents

37. Iron-core laminations are made from 37. ______________
    a. Silicon steel
    b. Hysteresis iron
    c. Pure iron
    d. Reluctance steel

NAME ____________________ DATE ____________

38. Eddy currents are reduced by **38.** ________
a. Laminating the core
b. Using a silicon-steel core
c. Using large-diameter wire in the secondary
d. Using large-diameter wire in the primary

39. Transformers have minimum efficiency when they are **39.** ________
a. Overloaded
b. Fully loaded
c. Half loaded
d. Unloaded

40. Decreasing the load on a transformer causes **40.** ________
a. The primary current to increase
b. The transformer to become less inductive
c. The power factor to increase
d. Angle θ to increase

41. Energizing current refers to the **41.** ________
a. Sum of the loaded primary and secondary currents
b. Fully loaded primary current
c. Unloaded primary current
d. Fully loaded secondary current

42. A rectifier transformer is a type of **42.** ________
a. Power transformer
b. Isolation transformer
c. Audio transformer
d. Constant-voltage transformer

43. In Fig. 12-2(*a*), if terminals 4 and 2 were connected, the two windings would be series-aiding. To connect the secondaries in parallel, the correct connections would be **43.** ________
a. 2 to 3 and 1 to 4
b. 3 to 4 and 2 to 1
c. 3 to 1 and 2 to 4

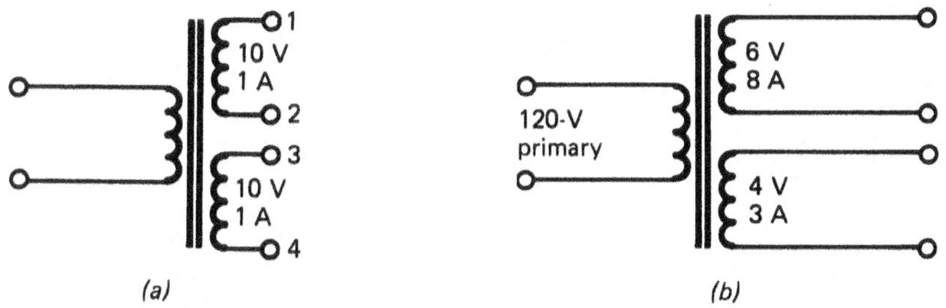

Fig. 12-2 Circuits for questions 43, 44, and 45.

44. Refer to Fig. 12-2(*b*). When connected series-opposing, the secondaries could provide **44.** ________
a. 10 V and 8 A
b. 10 V and 3 A
c. 10 V and 5 A
d. 2 V and 3 A

45. Refer to Fig. 12-2(*b*). When connected series-aiding, the secondaries could provide **45.** ________
a. 10 V and 8 V
b. 10 V and 3 A
c. 10 V and 5 A
d. 2 V and 3 A

46. Refer to Fig. 12-3. To form a wye primary, the terminals that could be connected together are
a. 1 to 3, 4 to 5, and 2 to 6
b. 1 to 4 to 6 and 2 to 3 to 5
c. 1 to 3 to 5
d. 2 to 3, 4 to 5, and 6 to 1

46. ______

47. Refer to Fig. 12-3. For a delta secondary, the terminals that could be connected together are
a. 8 to 10 to 12
b. 8 to 10 to 12 and 7 to 9 to 12
c. 8 to 9, 10 to 11, and 12 to 7
d. 12 to 10, 7 to 9, and 8 to 11

47. ______

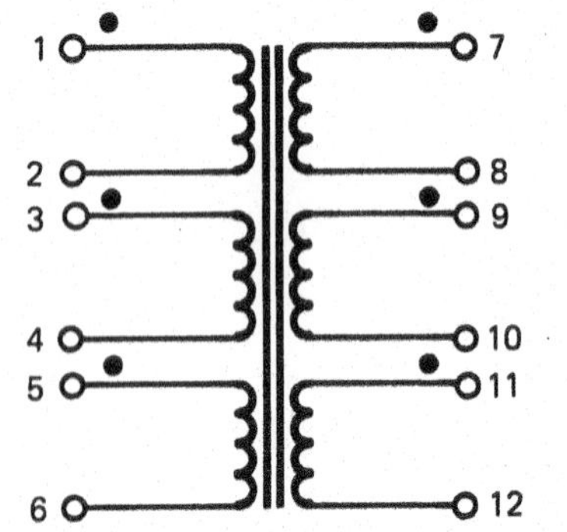

Fig. 12-3 Transformer for questions 46 and 47.

For questions 48 to 55, solve each problem. Be sure to include units (watts, degrees, etc.) in your answer when appropriate.

48. How much primary power is required by a transformer that is 92 percent efficient and provides 600 W of secondary power? 48. ______
49. What is the impedance ratio of a transformer that has 750 turns in the primary and 50 turns in the secondary? 49. ______
50. If the secondary load in question 49 is 14 Ω, what will the primary impedance be? 50. ______
51. The 208-V primary of a transformer has 1040 turns. How many turns does its 230-V secondary have? 51. ______
52. A multiple-secondary transformer has four turns per volt in its 20-V secondary. How many volts does its 50-turn secondary produce? 52. ______
53. What is the turns-per-volt ratio of the transformer in question 51? 53. ______
54. A transformer requires 850 W from its primary source to deliver 720 W to its load. What is its efficiency? 54. ______
55. A 100-Ω resistor is connected to a source that has 100 Ω of internal resistance. What is the efficiency? 55. ______

NAME ____________________ DATE ____________

12-1 LAB EXPERIMENT: TRANSFORMER EFFICIENCY AND POWER FACTOR

PURPOSE

This experiment will show you how to determine some of the characteristics of a transformer under load. You will gain additional experience in using a wattmeter and in constructing ac circuits.

MATERIALS

Qty.

1 transformer, 115/230-V primary, two 6-V, 2-A secondaries (Triad F-106Z or Stancor P-6376 or equivalent)
1 wattmeter, 50- or 100-W range, 150 V, 1 A
1 ammeter, ac, 300- or 500-mA range
1 VOM or DMM with ac voltage ranges
1 resistor, 6 Ω, 50 W (Dale RH-50)
1 resistor, 25 Ω, 12 W
1 0.5-A fuse (with holder)

CAUTION The resistor in this experiment can get very warm if the circuit is operated for extended periods of time. This circuit is powered from a 115-V ac outlet. Make all circuit connections before connecting to the 115-V outlet. After the circuit is connected to the outlet, do not touch any part of it with your fingers (or your body). When making voltage measurements, be certain that you touch only the insulated part of the test probes.

PROCEDURE

1. Construct the circuit shown in Fig. 12-4, but do not connect it to the 115-V source at this time. Refer back to experiment 9-1 if you have forgotten how to connect a wattmeter. Notice that the primary is parallel-connected and it operates from 115 V.
   a. If the primary were series-connected, how much voltage should it operate from to give rated values of output voltage? ____________
   b. Would the transformer operate equally well if terminals 1 and 4 and terminals 2 and 3 were connected to the 115-V source? ____________ Why?
   ____
   ____

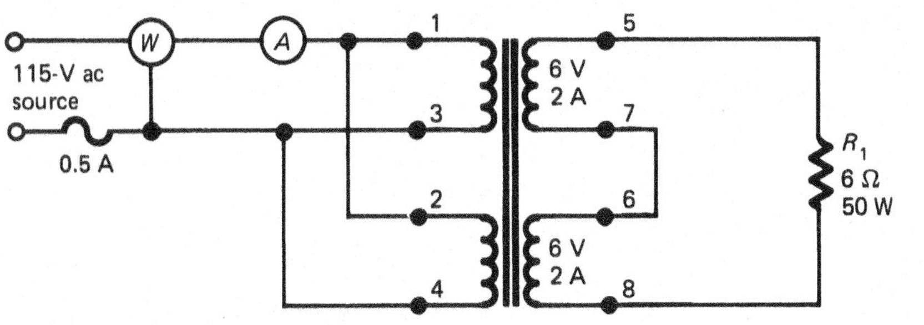

Fig. 12-4 Circuit for experiment 12-1.

2. Recheck your circuit to be certain the transformer terminals are connected as specified in Fig. 12-4. Now connect the circuit to the 115-V source.
   a. Record the power measured by the wattmeter. ______________
   b. Record the current measured by the anmeter. ______________
   c. Using your multimeter on the 15-V ac range, measure and record the voltage across the 6-Ω resistor. Disconnect (unplug) the circuit from the 115-V source. You now have the data needed to determine efficiency, power factor, and angle θ. (This assumes that the outlet voltage in your laboratory is actually 115 V. If this point is in doubt, measure and record the actual value for use in all the calculations to follow.) ______________
3. a. Calculate and record the apparent power drawn from the primary source. (Use your measured value of current and the source voltage.) ______________
   b. Next, use the apparent power just calculated and the true power (measured) to determine the power factor (cos θ). ______________
   c. From the calculated power factor (cos θ), determine the angle θ by which primary current and voltage are out of phase. ______________
   d. Do your results in this section indicate that a loaded transformer appears to the source to be mostly inductive or resistive? ______________
4. a. To determine efficiency, you must first calculate the power delivered by the secondary. You can do this in two steps with

$$I = \frac{V}{R} \text{ and } P = VI$$

   Or, you can do it in one step with

$$P = \frac{V^2}{R}$$

   (Use the voltage you measured across R_1 and the value of R_1 shown in Fig. 12-4.) P_{load} = ______________
   b. Using this calculated value of secondary power and the measured value of primary power from step 2a, determine and record the transformer's efficiency. ______________
5. Change the 6-Ω resistor in Fig. 12-4 to a 25-Ω resistor. This decreases the load on the transformer to about 25 percent of its former value. Connect the circuit to the 115-V source, and record the new current and power indicated by the meters. Measure and record the voltage across R_1. Disconnect the power from the 115-V source. Compare the current and power just recorded with the values recorded in steps 2a and b.
   P_{pri} = ______________
   I_{pri} = ______________
   V_{R_1} = ______________
   a. Did the primary current drop as much, proportionately, as the power? ______________
   b. Did the apparent power decrease as much, proportionately, as the true power? ______________
   c. Did the power factor increase or decrease? ______________
   d. Did the primary power decrease as much, proportionately, as the secondary power? ______________
   e. Did the efficiency increase or decrease? ______________

NAME ______________________________ DATE ______________

12-2 LAB EXPERIMENT: SERIES AND PARALLEL WINDINGS

PURPOSE

This experiment will provide experience in proper phasing of windings for both series and parallel connections. It will show you how to adjust secondary voltage by aiding or bucking a small part of the primary winding.

MATERIALS

Qty.

1 transformer, 115/230-V primary, two 6-V, 2-A secondaries (Triad F-106Z or Stancor P-6376 or equivalent)

Qty.

1 fuse, 0.5 A (with holder)
1 VOM or DMM

CAUTION **This circuit is powered from a 115-V ac outlet. Make all circuit connections before connecting to the 115-V outlet. After the circuit is connected to the outlet, do not touch any part of it with your fingers (or your body). When making voltage measurements, be certain you touch only the insulated part of the test probes.**

PROCEDURE

1. The start end of each winding of the transformer is indicated in Fig. 12-5.

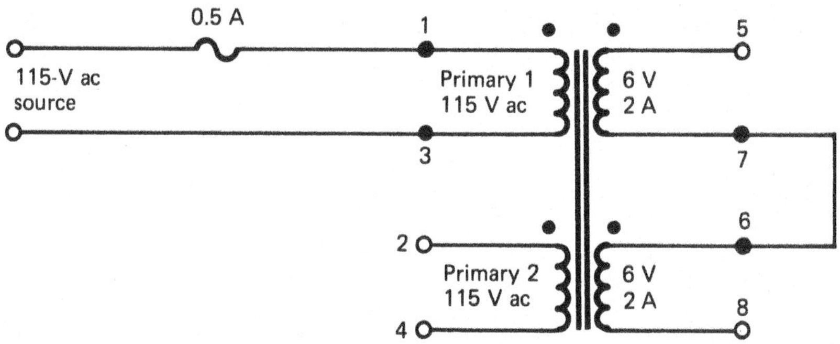

Fig. 12-5 Circuit for experiment 12-2, steps 1 through 4.

a. Will connecting terminals 7 and 6 (7–6) together produce series-aiding or series-opposing secondaries? ______________

b. Will the output voltage be approximately 0, 6, or 12 V?

____

c. Construct the circuit in Fig. 12-5 to see whether your answers to the questions in steps 1a and b are correct. We are using only one of the primary windings. Since we will not be loading the secondaries, one primary can easily carry the required primary current. (There is no need to connect the primaries in parallel and complicate the circuit wiring.) Now connect the circuit of Fig. 12-5 to the 115-V source. Measure and record the voltage available from the series-connected secondaries (measure between terminals 5 and 8). ______________

d. Disconnect the circuit from the 115-V source. Does the measured secondary voltage verify your answers to questions in steps 1a and b?

____

2. a. Remove the connection between terminals 6 and 7 in Fig. 12-5 and connect terminals 5 and 8 together. What type (series-aiding or series-opposing) of connection will this provide? ______________

b. Check your answer by applying power (connecting to 115-V source), and measuring the output voltage (voltage between terminals 6 and 7).

____

c. Remove power from the circuit. Does the measured voltage support your answer to step 2a? ______________

3. a. Remove the connection between terminals 5 and 8, and connect terminals 5 and 6 together. How much output voltage do you anticipate now? ______________

b. Apply power to the circuit and measure the voltage between terminals 7 and 8. Remove power from the circuit. Are the secondaries now connected series-aiding or series-opposing? ______________

4. When terminals 5 and 6 are connected together, there is no voltage between terminals 7 and 8. (This was shown in step 3.) Therefore, it is safe to connect terminals 7 and 8 together while 5 and 6 are still connected.

a. Make the connection between terminals 7 and 8, and apply power to the circuit. Measure and record the output voltage between terminals 5 and 8. ______________

b. Remove power from the circuit. Are the secondary windings now connected in series or in parallel? ______________

c. Are there any other combinations of terminals that yield correctly phased parallel secondaries? ______________

5. a. Refer to Fig. 12-6(*a*). Are the primaries connected series-aiding or series-opposing? ______________

b. If they are series-aiding and the primary source voltage has not been doubled, what will happen to the secondary voltage? ______________

c. Construct the circuit of Fig. 12-6(*a*) and apply power. Measure and record the voltage of either secondary. ______________

d. Now, disconnect the primary from the 115-V source. Is the turns-per-volt ratio of the transformer as connected in Fig. 12-6(*a*) the same as it was in Fig. 12-5? ______________

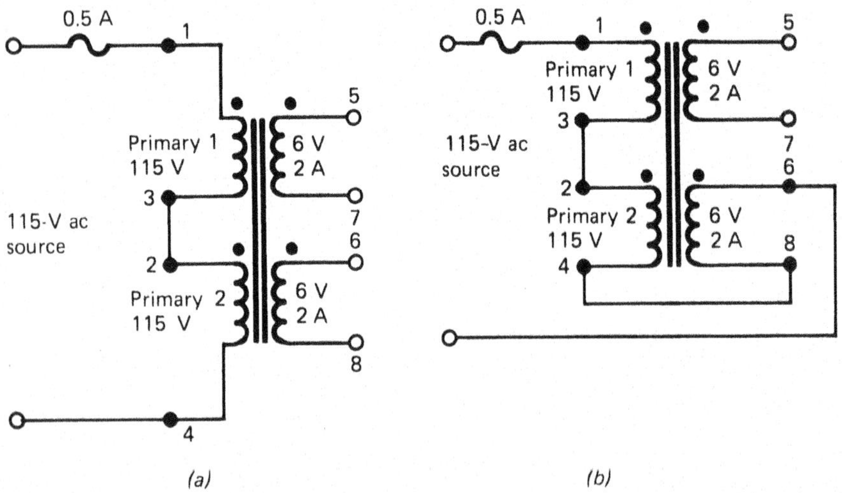

Fig. 12-6 Circuits for experiment 12-2, steps 5 and 6.

6. In this part of the experiment we will use one of the 6-V secondaries as part of the primary. The turns of the 6-V winding (between terminals 6 and 8) can be connected either aiding or opposing the rest of the primary.

a. Construct the circuit shown in Fig. 12-6(*b*) and apply power. Measure and record the secondary voltage (between terminals 5 and 7). Remove the power from the circuit. ______________

b. Is the voltage found in step 6a less or more than that measured in step 5c? ______________

c. Therefore, is the additional primary aiding or opposing the two original primaries? ______________

d. Be sure the circuit is disconnected from the 115-V source. Now reverse the connections to terminals 6 and 8. Will this new connection increase or decrease the secondary voltage? ______________

e. Apply power to the circuit, and check your answer by measuring and recording the secondary voltage. Remove power from the circuit. ______________

12-3 LAB EXPERIMENT: IMPEDANCE MATCHING

PURPOSE

This experiment has two objectives. The first objective is to verify that maximum power is transferred when impedances are matched. The second objective is to show that a transformer can match a small load impedance to a large source impedance (or vice versa).

MATERIALS

Qty.

1 transformer, 115/230-V primary, two 6-V, 2-A secondaries (Triad F-106Z or Stancor P-6376 or equivalent)
1 0.5-A fuse (with holder)
1 transformer, approximately 10:1 turns ratio (Triad F-106Z or Stancor P-6376 or equivalent)
1 resistor, 2.2 Ω, 5%, ½ W
1 resistor, 4.7 Ω, 5%, ½ W
1 resistor, 10 Ω, 5%, ½ W
1 resistor, 22 Ω, 5%, ½ W
1 resistor, 47 Ω, 5%, ½ W
1 resistor, 220 Ω, 5%, ½ W
1 resistor, 470 Ω, 5%, ½ W
2 resistors, 1000 Ω, 5%, ½ W

Qty.

1 resistor, 4700 Ω, 5%, ½ W
1 DMM or VOM with a low (less than 2 V) ac voltage range
Multisim
Multisim file for Fig. 12-7(*b*)
Multisim file for Fig. 12-9(*a*)

INTRODUCTION

In this experiment we will use the power from a transformer secondary as the source. The transformer used has a very low internal impedance, so we will add a series resistor and consider it part of the internal resistance. For this experiment then, consider the source to be everything inside the broken lines in Fig. 12-7(*a*). The 1000 Ω of resistance [R_1 in Fig. 12-7(*a*)] is very high compared with the rest of the impedance of the transformer. Therefore, for this experiment, we can consider the source impedance to be 1000 Ω. Figure 12-7(*b*) shows the equivalent circuit for Fig. 12-7(*a*). The load impedances will consist of a number of resistors.

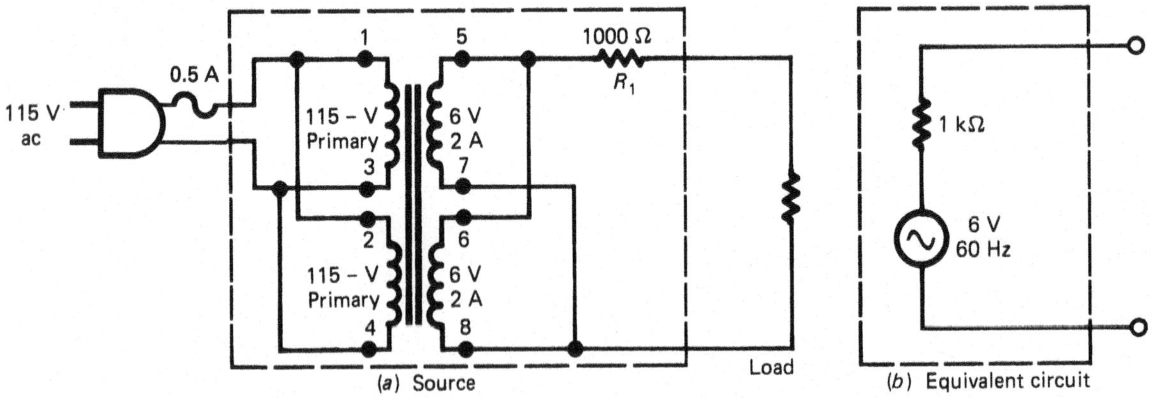

Fig. 12-7 An ac source with 1000 Ω of internal resistance.

In an electronic circuit the high-impedance source could be the output of an amplifier or a microphone. The load might be a speaker or the low input impedance of another amplifier.

If you are using Multisim for this experiment, you will be using the Multisim file for Fig. 12-7(*b*) in step 1 and the Multisim file for Fig. 12-9(*a*) in step 2.

CAUTION **This circuit is powered from a 115-V ac outlet. Make all circuit connections before connecting to the 115-V outlet. After the circuit is connected to the outlet, do not touch any part of it with your fingers (or your body). When making voltage measurements, be certain that you touch only the insulated part of the test probes.**

PROCEDURE

1. Using a 220-Ω resistor as the load, construct the circuit of Fig. 12-7. Apply power to the circuit. Measure the load voltage and record it in the table in Fig. 12-8(*a*). Now calculate the power used by the load and record it in Fig. 12-8(*a*).

Without impedance-matching transformer

Load resistance (ohms)	Load voltage measured (volts)	Load power calculated (watts) $P = V^2/R$
220		
470		
1000		
2200		
4700		

(*a*)

With impedance-matching transformer

Load resistance (ohms)	Load voltage measured (volts)	Load power calculated (watts) $P = V^2/R$
2.2		
4.7		
10		
22		
47		

(*b*)

Fig. 12-8 Data tables for experiment 12-3.

a. Disconnect the power, change the load to 470 Ω, and reapply the power. Measure and record the new load voltage. Also, calculate and record the new load power. Did the load power increase with the high resistance? ______________

b. With which of the load resistances listed in Fig. 12-8(*a*) do you expect the greatest load power? ______________

c. Repeat the procedure outlined in step 1a until the table in Fig. 12-8(*a*) is completed. Do your results support your answer to step 1b? ______________

d. Do your results support the statement "Maximum power is transferred when the source and the load impedances are matched"? ______________

2. Look at Fig. 12-9. It is like Fig. 12-7 except that it has an impedance-matching transformer between the source and the load. The secondaries are connected series-aiding. Thus, the voltage ratio of the transformer is 115 to 12, or approximately 10:1.

a. What is the turns ratio of the transformer? ______________

b. What is the impedance ratio of the transformer? ______________

c. Which of the resistances listed in Fig. 12-8(*b*) should provide the correct reflected impedance to match the source impedance?

____

d. Construct the circuit shown in Fig. 12-9 (start with the 2.2-Ω load), and apply power to the circuit. Since the matching transformer steps down the voltage, the load voltage will be quite small (less than 1 V). Using the lowest ac voltage range of your VOM or DMM, measure and record in Fig. 12-8(*b*) the load voltage of Fig. 12-9. Then calculate the load power, and record it in the data table of Fig. 12-8(*b*). Change the load resistor to the next higher value and again measure load voltage and calculate load power. Continue this process until the table is completed. Do your results support your answer to the question in step 2c?

____

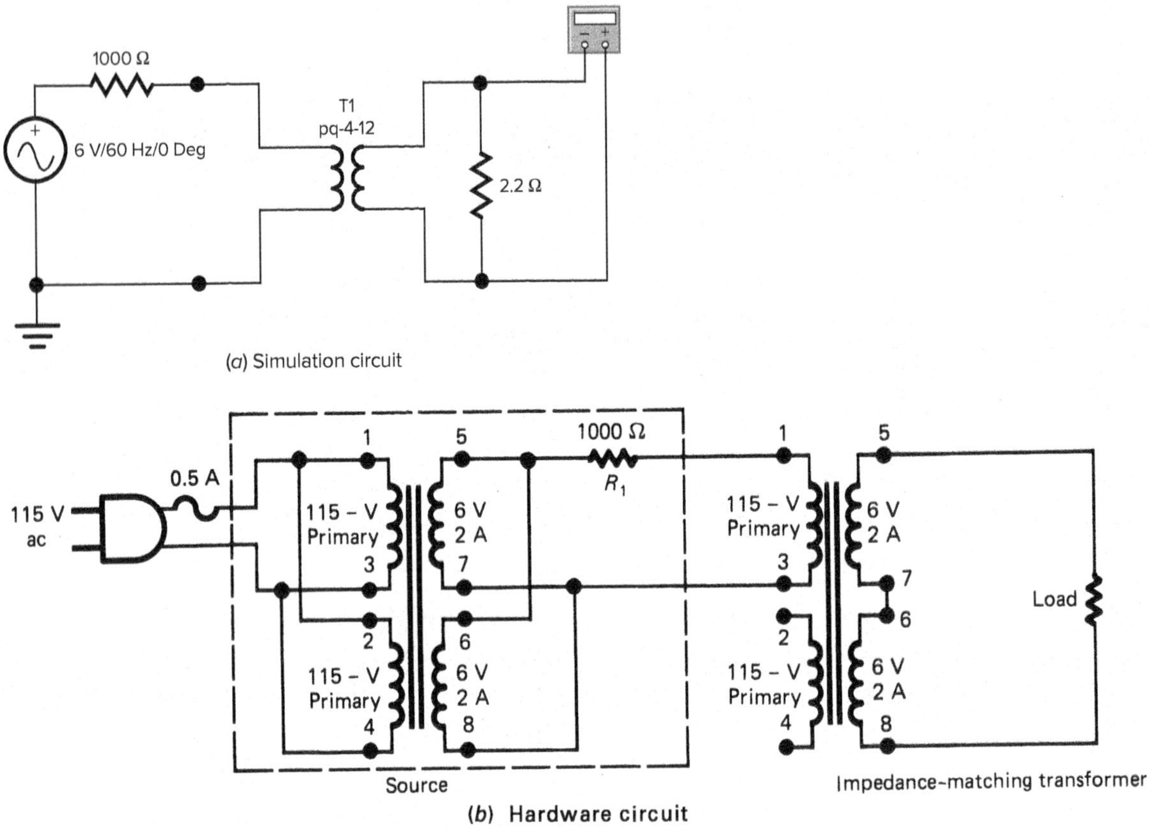

Fig. 12-9 Using an impedance-matching transformer.

e. Did the matching transformer make the load resistors appear to the source to be higher or lower than their true value? ______________ How much higher or lower?

____

____

f. Notice that the maximum power reaching the load was less in Fig. 12-8(*b*) than in Fig. 12-8(*a*). What do you think would cause this?

____

____

g. What value of load resistance would match the source if we used only one secondary of the matching transformer? ______________

h. What value of load resistance would match the source if we used both primaries (series-aiding) and both secondaries (series-aiding) of the matching transformer? ______________

NAME ____ DATE ______________

12-4 LAB EXPERIMENT: THREE-PHASE TRANSFORMER CONNECTIONS

PURPOSE

This experiment provides experience in connecting three single-phase transformers for transforming three-phase voltages. It will illustrate the differences between wye and delta connections and between balanced and unbalanced loads.

MATERIALS

Qty.		Qty.	
3	transformers, single-phase, 230-V primary, 6-V, 2-A secondary (Triad F-106Z or Stancor P-6376 or equivalent)	3	no. 47 lamps, 6.3 V, 150 mA, with holders
3	0.5-A fuses (with holders)	2	ammeters, ac, 300- or 500-mA range
		1	VOM or DMM

CAUTION **A shock received from a 208-V circuit is very dangerous. It can be fatal. Make all circuit connections before applying power (connecting to the three-phase source). Once power is applied, do not touch any part of the primary circuit.**

PROCEDURE

1. Study the circuit in Fig. 12-10. It uses three transformers to make the equivalent of a single three-phase transformer. Notice that the two primary windings of each transformer are connected series-aiding so they can operate on 208 V (although rated at 230 V). One secondary (terminals 6 and 8) of each transformer is not used. They have been left off the schematic diagram to keep the diagram from getting too cluttered.

   Construct the circuit in Fig. 12-10. Check your circuit to be sure that the terminal connections are as shown on the diagram. Then apply power.

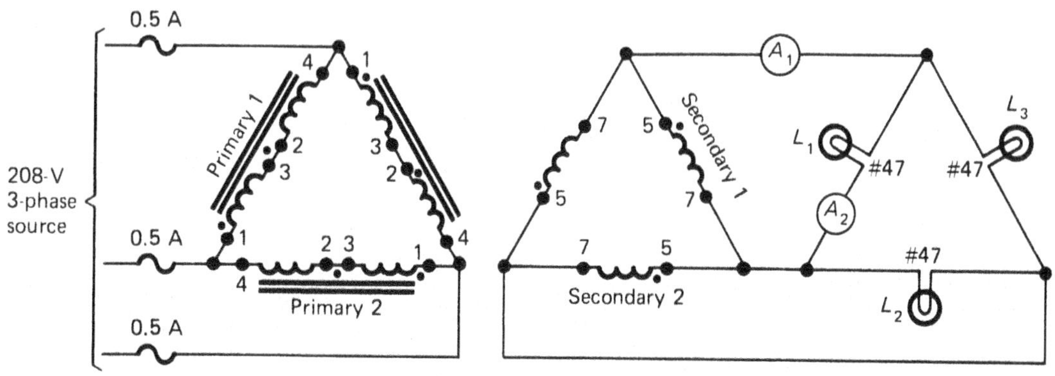

Fig. 12-10 Delta-delta connection connected to a balanced load.

   a. Do all three lamps light with equal brightness? ______________
   b. Is the line current (meter A_1) or the load current (meter A_2) the higher?
   c. Is the higher current approximately 1.73 times as high as the lower current? ______________

d. Remove L_3 from its socket. Do the other two lamps still light with full brightness? ______________

e. Did the current through A_1 increase, decrease, or stay the same? ______________

f. Did the current through A_2 increase, decrease, or stay the same? ______________

Now you are going to check the phasing of the delta-connected secondary. **Disconnect the circuit from the power source.** Remove the loads and ammeters. Interrupt the delta and insert the VOM (20-V ac range) as shown in Fig. 12-11(*a*). (The primary connections are not shown in Fig. 12-11. They are the same as before.) Apply power to the circuit.

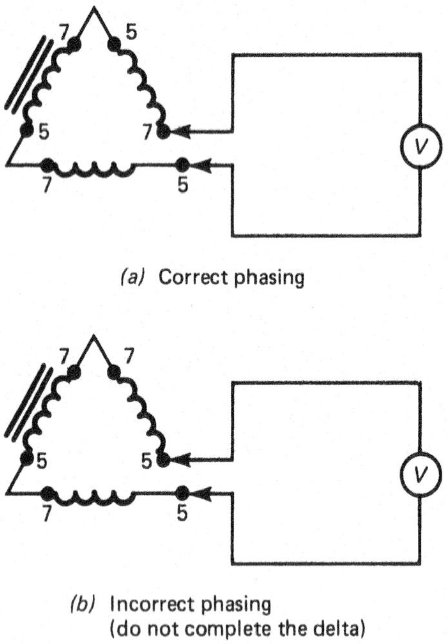

Fig. 12-11 Checking delta phasing on a transformer secondary.

g. How much voltage does the meter indicate? ______________

h. Remove power from the circuit and reverse the leads to one of the secondaries [Figure 12-11(*b*)] to provide incorrect phasing. Again apply power to the circuit. How much voltage does the meter indicate now? Remove power from the circuit. ______________

CAUTION **A shock received from a 208-V circuit is very dangerous. It can be fatal. Make all circuit connections before applying power (connecting to the three-phase source). Once power is applied, do not touch any part of the primary circuit. Disconnect the power source before changing circuit connections.**

2. Notice that the circuit in Fig. 12-12 uses the same primary connections as you have been using. However, it has a wye-connected secondary with a balanced load. Construct the circuit in Fig. 12-12 and apply power.

a. Do all the lamps light with equal brightness? ______________

b. Measure the line voltages (line 1 to line 2, line 2 to line 3, and line 1 to line 3). Are all the line voltages the same? ______________

c. Measure the load voltages across the individual lamps. Are they all equal? ______________

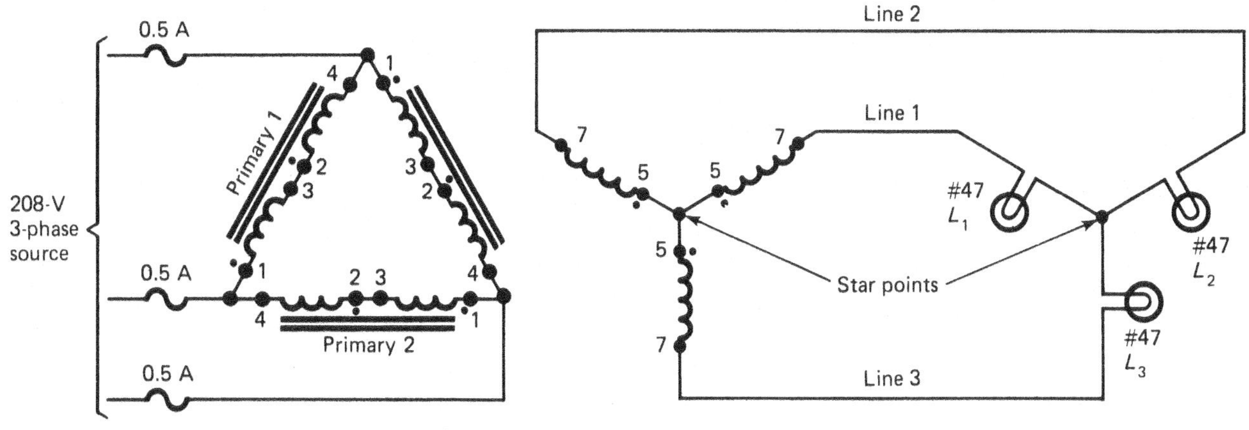

Fig. 12-12 Delta-wye connection connected to a balanced load.

d. Is the line voltage approximately 1.73 times the load voltage? ______________

e. The circuit in Fig. 12-12 has a balanced load. Should there be any voltage between the star point of the load and the star point of the secondaries? ______________

f. Connect a 300-mA ac meter between the star points. How much current flow does the meter indicate? ______________

g. If the loads are perfectly balanced and the secondaries are identical, should there be any current indicated on the meter? Remove the ammeter from between the star points. ______________

h. Now remove any one of the three lamps from its socket. The wye now has an unbalanced load. Do the two remaining lamps light as brightly as they did when all three lamps were present? ______________

i. Connect a conductor (called a neutral wire) between the two star points. Do the two remaining lamps light to full brightness now? ______________

j. Remove the neutral wire, and connect the ammeter between the star points. How much current does it indicate? ______________

k. Disconnect power from the circuit. Does a neutral wire in a three-phase wye carry current when the loads are unbalanced? ______________

3. Be sure the power is removed from the circuit. Remove all loads and the ammeter from the circuit. Now reverse the secondary terminal connections on one transformer to provide the secondary configuration in Fig. 12-13. Figure 12-13 is an improperly phased wye connection (the primary connections, not shown, are the same as in Figs. 12-10 and 12-12). Apply power to the incorrectly phased circuit.

a. Measure all three of the phase voltages (star point to terminal 5 or 7). Are they all equal? ______________

b. Measure the three line voltages (terminals 7 to 7, 7 to 5, and 5 to 7). Are they all equal? ______________

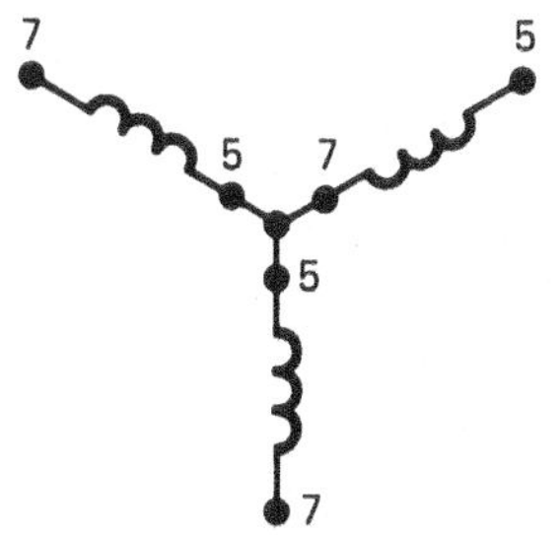

Fig. 12-13 Incorrect phasing of a wye-connected transformer secondary.

12-7 LAB EXPERIMENT: THREE-PHASE CONNECTIONS

PURPOSE

This experiment will provide experience in designing, constructing, and testing a three-phase system with multiple loads.

MATERIALS

Qty.

Multisim
Multisim file for Fig. 12-16

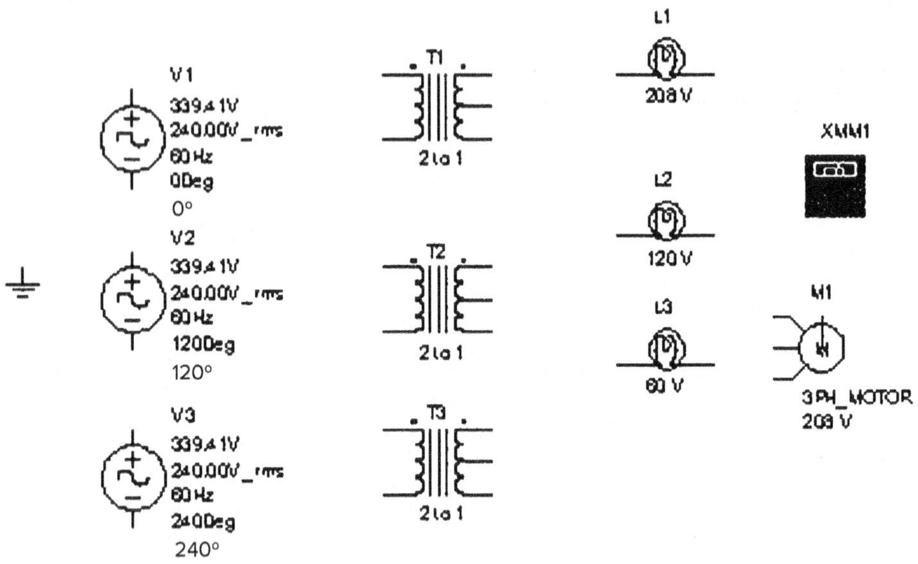

Fig. 12-15 Components available for experiment 12-7.

PROCEDURE

1. Using Multisim, open the file for Fig. 12-16.
2. Using only the components and equipment shown in the figure, design/connect a three-phase system that will provide the correct voltages to all four loads.
3. When you have a working circuit, demonstrate it to your instructor.

NAME ____________________ DATE ____________________

CHAPTER 13

R, *C*, and *L* Circuits

TEST: *R*, *C*, AND *L* CIRCUITS

For questions 1 to 27, determine whether each statement is true or false.

Statement	Answer
1. Current and voltage are always out of phase in a circuit that has only resistance and inductance.	**1.** ________
2. Current leads voltage in an *RC* circuit.	**2.** ________
3. The Pythagorean theorem can be used to find the resultant of a series-connected resistance and reactance.	**3.** ________
4. Impedance and reactance are never 90° out of phase.	**4.** ________
5. Impedance has no base unit.	**5.** ________
6. A resistance of 6 Ω in series with 8 Ω of reactance results in an impedance of 14 Ω.	**6.** ________
7. In a series circuit, if the value of *R* is one-half that of *X*, then V_R is one-half of V_X.	**7.** ________
8. Increasing the resistance of a series *RC* circuit causes angle θ to increase.	**8.** ________
9. In a parallel *RL* circuit, both *R* and X_L must be larger than the impedance.	**9.** ________
10. In a phasor diagram for a series circuit, the current phasor is the reference phasor.	**10.** ________
11. The capacitive voltage can be higher than the source voltage in a series *RCL* circuit.	**11.** ________
12. The impedance may be lower than the resistance in a series *RCL* circuit.	**12.** ________
13. The impedance of a parallel resonant circuit is higher than the impedance of a series resonant circuit.	**13.** ________
14. A series *RCL* circuit is inductive when X_L is greater than X_C.	**14.** ________
15. A parallel *RCL* circuit is capacitive when X_C is greater than X_L.	**15.** ________
16. Inductive reactance equals capacitive reactance for both series and parallel resonance.	**16.** ________
17. Decreasing *C* increases the resonant frequency of a parallel *LC* circuit.	**17.** ________
18. Increasing *L* increases the resonant frequency of a series *LC* circuit.	**18.** ________
19. A 0.01-μF and a 1-mH inductor produce parallel resonance at more than one frequency.	**19.** ________
20. For both parallel and series resonant circuits, angle θ is zero.	**20.** ________
21. For a parallel *RC* circuit, cos θ is equal to the resistive current divided by the source voltage.	**21.** ________
22. A parallel resonant circuit is commonly referred to as a tank circuit.	**22.** ________
23. The quality *Q* of a resonant circuit is determined primarily by the *Q* of the capacitor.	**23.** ________
24. A series resonant *LC* circuit draws a high current from its source.	**24.** ________
25. A high-*Q* circuit has a wider bandwidth than a low-*Q* circuit does.	**25.** ________
26. A high-pass filter passes high frequencies around the load.	**26.** ________

27. A series resonant circuit can be used in either a band-pass or a band-reject filter. 27. ____________

For questions 28 to 35, choose the letter that best completes each statement.

28. The symbol or abbreviation for impedance is 28. ____________
  a. I
  b. Q
  c. X
  d. Z

29. The formula that would not be used to determine the power factor in a series *RC* circuit is 29. ____________
  a. $\cos\theta = \frac{V_R}{V_T}$
  b. $\cos\theta = \frac{I_R}{I_T}$
  c. $\cos\theta = \frac{P}{P_{\text{app}}}$
  d. $\cos\theta = \frac{X_C}{Z}$

30. Impedance is always greater than resistance in 30. ____________
  a. Parallel *RC* circuits
  b. Parallel *RL* circuits
  c. Series *RL* circuits
  d. Series *RCL* circuits

31. Increasing the frequency decreases the power in a 31. ____________
  a. Parallel *RC* circuit
  b. Parallel *RL* circuit
  c. Series *RC* circuit
  d. Series *RL* circuit

32. Reactive voltage can be greater than source voltage in a 32. ____________
  a. Series *RL* circuit
  b. Series *RCL* circuit
  c. Parallel *RC* circuit
  d. Parallel *RCL* circuit

33. Reactive current can be greater than source current in a 33. ____________
  a. Series *RL* circuit
  b. Series *RCL* circuit
  c. Parallel *RC* circuit
  d. Parallel *RCL* circuit

34. When a circuit is resonant, 34. ____________
  a. $L = C$
  b. $X_L = X_C$
  c. $\theta = 45°$
  d. $Q = Z$

35. To be most selective, a parallel resonant *LC* circuit should have a 35. ____________
  a. High-quality inductor
  b. Low-quality capacitor
  c. Cos θ equal to zero
  d. Large bandwidth

For questions 36 to 47, solve each problem. Be sure to include units (watts, degrees, etc.) when appropriate.

36. A parallel *LC* circuit is resonant at 150 kHz. It uses a 2.5-mH inductor with 25 Ω of resistance. What is the quality of the circuit? 36. ____________

37. What is the quality of a circuit that is resonant at 4 MHz and has a bandwidth of 30 kHz? 37. ____________

38. What is the bandwidth of the circuit in question 36? 38. ____________

NAME ______________________ DATE ______________

39. An 1800-Ω resistor is in series with a 0.15-μF capacitor. They are connected to a 20-V, 350-Hz source. Determine the following:
a. Impedance
b. Source current
c. Power
d. Power factor

39. *Place answers below.*
a. ______
b. ______
c. ______
d. ______

40. What is the resonant frequency of a 0.1-mH inductor and a 560-pF capacitor?

40. ______

41. Suppose the components in question 39 were connected in parallel rather than in series. Determine the following:
a. Impedance
b. Capacitive current
c. Theta
d. Source current

41. *Place answers below.*
a. ______
b. ______
c. ______
d. ______

42. For the circuit in Fig. 13-1, do the following:
a. Determine the impedance.
b. Determine the power factor (cos θ).
c. Determine the voltage across *C*.
d. Determine whether the circuit is inductive or capacitive.

42. *Place answers below.*
a. ______
b. ______
c. ______
d. ______

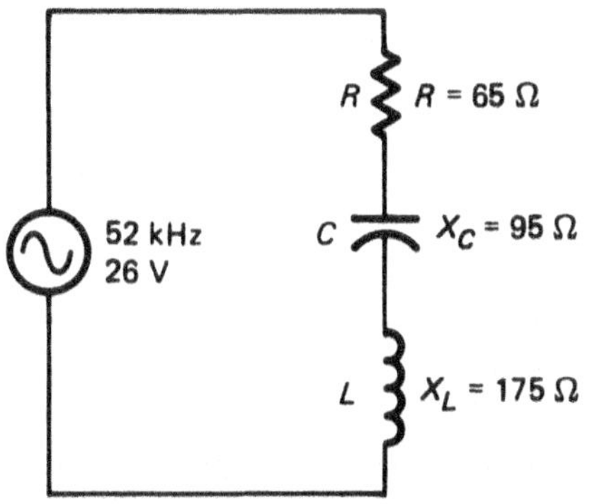

Fig. 13-1 Circuit for question 42.

43. For the circuit in Fig. 13-2, do the following:
a. Compute the current through *L*.
b. Compute the impedance.
c. Determine whether the circuit is inductive or capacitive.
d. Compute the power factor.

43. *Place answers below.*
a. ______
b. ______
c. ______
d. ______

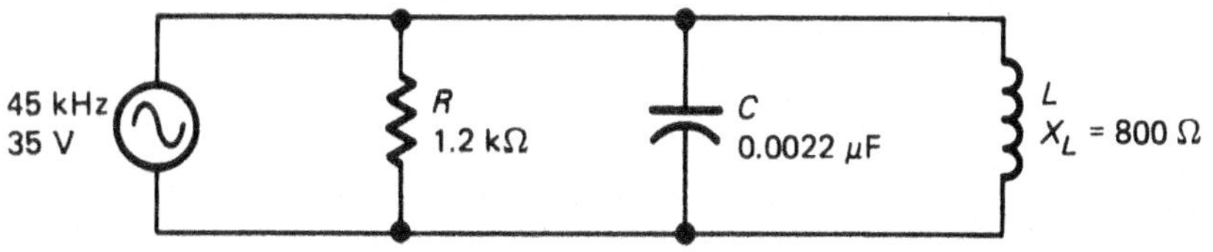

Fig. 13-2 Circuit for question 43.

44. In a series *RC* circuit, the resistance is 1000 Ω, the capacitance is 0.15 μF, and the frequency is 500 Hz.
Determine
a. Impedance
b. Power factor

44. *Place answers below.*
a. ______
b. ______

45. What is the impedance of a parallel RC circuit when $R = 2000\ \Omega$, $C = 0.68\ \mu F$, and $f = 150$ Hz?

45. ____________________

46. For a series RCL circuit, $R = 56\ \Omega$, $X_C = 80\ \Omega$, and $X_L = 60\ \Omega$. Determine

a. Impedance

b. Power factor

46. *Place answers below.*

a. ____________________

b. ____________________

47. For the circuit in question 46 when it is connected to a 20-V source, determine

a. I_T

b. P

47. *Place answers below.*

a. ____________________

b. ____________________

NAME ______ DATE ______

13-1 LAB EXPERIMENT: IMPEDANCE

PURPOSE

This experiment will provide experience in making and interpreting measurements in *R*, *C*, and *L* combination circuits. It will also verify many of the formulas used in solving impedance problems.

MATERIALS

Qty.		Qty.	
1	signal generator, sinusoidal, audio to 150-kHz range	1	resistor, 1000 Ω, ±10%, ½ W
1	VOM or DMM	1	resistor, 2200 Ω, ±10%, ½ W
1	ammeter, ac, 0 to 5 mA, ± 3% accuracy or equivalent	1	capacitor, 1500 pF, ±10%, 100 WV dc (minimum)
1	choke, 2.5 mH, 125 mA, 38-Ω dc resistance	1	capacitor, 0.47 μF, ±10%, 100 WV dc (minimum)
1	resistor, 470 Ω, ±10%, ½ W	1	capacitor, 1 μF, ±10%, 100 WV dc (minimum)

PROCEDURE

1. The circuit in Fig. 13-3 is a series *RC* circuit operating at 160 Hz. You will determine its impedance, current, and voltage drops both mathematically and by measurement. First, solve the circuit mathematically (assume the meter has zero internal resistance).

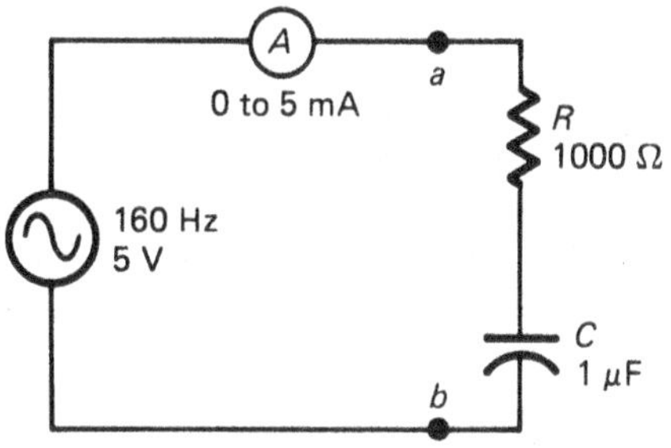

Fig. 13-3 Circuit for experiment 13-1, steps 1, 2, and 3.

   a. What is X_C? ______
   b. What is *Z*? ______
   c. What is I_T? ______
   d. What is V_R? ______
   e. What is V_C? ______
   f. Now construct the circuit in Fig. 13-3. Be sure the generator is set for 160 Hz. Connect the DMM (ac voltage range) between points *a* and *b* and adjust the generator's voltage output until the DMM indicates 5 V. How much current does the milliammeter indicate? ______
   g. Does the measured current agree, within 10 percent, with the calculated current? ______
   h. Next, using the DMM, measure and record the voltages across the resistor and the capacitor.
   V_R ______
   V_C ______
   i. Do these voltages agree, within 10 percent, with the calculated values in steps 1d and e? ______

j. Using the current measured in step 1f and the 5-V source voltage between points a and b, calculate and record the impedance. _______________

k. Does this value of impedance agree, within 10 percent, with the value determined in step 1b? _______________

2. If the frequency in Fig. 13-3 is decreased to 100 Hz, will the following quantities increase or decrease?

a. I_T _______________

b. V_R _______________

c. V_C _______________

Change the frequency of the generator to 100 Hz and adjust the output for 5 V between points a and b. Now measure and record the following current and voltages at this new frequency:

d. I_T _______________

e. V_R _______________

f. V_C _______________

g. Compare the current recorded in step 2d with that in step 1f. Was your response to step 2a correct? _______________

Check your response to steps 2b and c by comparing the voltages measured in step 1 with those measured in step 2. Reset the generator frequency control for 160 Hz.

3. Suppose the capacitor in Fig. 13-3 were changed to 0.47 μF. Would the following quantities increase or decrease?

a. I_T _______________

b. V_R _______________

c. V_C _______________

Change the capacitor in Fig. 13-3 to a 0.47-μF capacitor. Adjust the voltage output for 5 V between points a and b (be sure the frequency is set for 160 Hz). Now measure and record the following current and voltages:

d. I_T _______________

e. V_R _______________

f. V_C _______________

g. Compare these voltages and currents with those measured in step 1. Were your predictions in steps 3a, b, and c correct? _______________

4. Refer to the parallel RC circuit in Fig. 13-4(a). Calculate and record the following reactance, currents, and impedance for the circuit:

a. X_C _______________

b. I_C _______________

c. I_R _______________

d. I_T _______________

e. Z _______________

f. Now construct the circuit shown in Fig. 13-4(a). Adjust the generator's voltage control to provide 4 V between points a and b. Record the current indicated on the meter. _______________

g. With this measured current and the 4 V of the source, determine the impedance. _______________

h. Are the values determined in steps 4f and g within 10 percent of the values calculated for steps 4d and e? _______________

i. Now change the location of the ammeter to that in Fig. 13-4(b). Measure and record I_C. _______________

j. Next change the ammeter to the location shown in Fig. 13-4(c). Measure and record I_R. _______________

k. Do these measured currents agree, within 10 percent, with the currents recorded in steps 4b and c? _______________ They should!

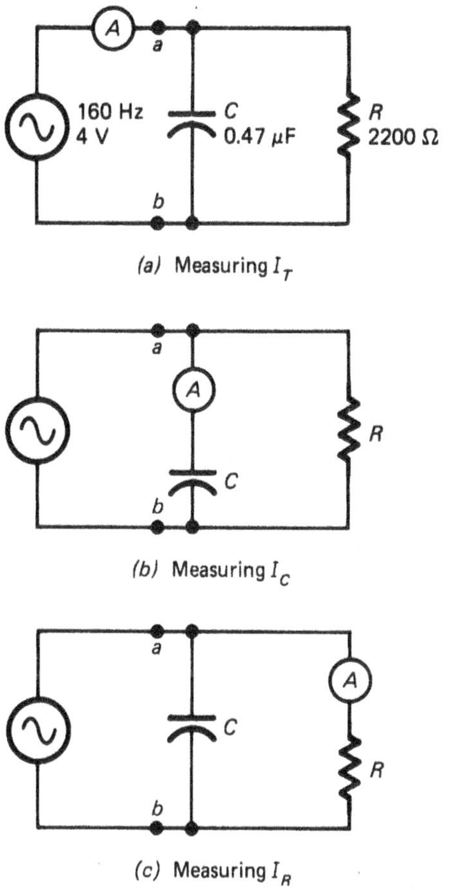

Fig. 13-4 Circuits for experiment 13-1, steps 4 and 5.

5. If you decrease the frequency in the three circuits in Fig. 13-4, what will happen to each of the following (answer "increase," "decrease," or "no change")?

a. I_R ________________

b. I_C ________________

c. I_T ________________

Check your responses by changing the frequency to 100 Hz and measuring these three currents:

d. I_R ________________

e. I_C ________________

f. I_T ________________

g. What happened to the phase angle (angle θ) when the frequency was decreased from 160 to 100 Hz? ________________

6. Construct the circuit shown in Fig. 13-5. Be sure the generator controls are set for 100 kHz. Adjust the voltage controls for 5 V across the circuit. Measure and record the following voltages:

a. V_L ________________

b. V_C ________________

c. V_R ________________

Recall two things about series circuits. One is that the current throughout the circuit is the same. The other is that the individual voltage drops are proportional to the individual oppositions. From your measured voltages,

d. is X_L greater or less than X_C? ________________

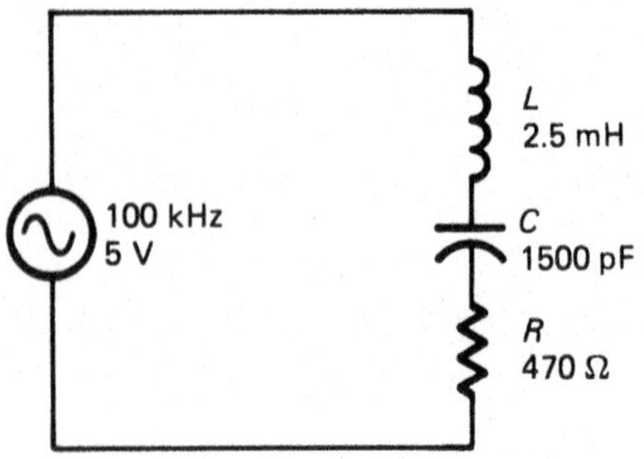

Fig. 13-5 Circuit for experiment 13-1, steps 6 and 7.

e. is X_C greater or less than R? ______________
f. is X_C greater or less than Z? ______________
g. is R greater or less than Z? ______________
h. Now, using Ohm's law and the measured resistive voltage, calculate the current in the circuit ($I_T = I_R = V_R/R$). ______________
i. Using this current and the source voltage (5 V), calculate the impedance of the circuit. ______________
j. Finally, using $X_C = V_C/I_C$, calculate the reactance of the capacitor. Do your answers to steps 6i and j support your answers to steps 6f and g? If not, you have made an error. ______________

7. If the frequency in Fig. 13-5 is doubled, will the following increase or decrease?

a. V_L ______________
b. V_C ______________
c. V_R ______________
d. I_T ______________

Check your answers by changing the frequency and making the appropriate measurements.

NAME ______________________________ DATE ______________

13-2 LAB EXPERIMENT: RESONANCE, QUALITY, AND BANDWIDTH

PURPOSE

The purpose of this experiment is to provide experience in measuring and determining the characteristics of resonant circuits.

MATERIALS

Qty.		Qty.	
1	signal generator, sinusoidal, audio to 150-kHz range	1	capacitor, 1000 pF, ±10%, 100 WV dc (minimum)
1	choke, 2.5 mH, 125 mA, 38-Ω dc resistance	1	resistor, 1 kΩ, ±10%, ½ W
		1	multimeter, 10-MΩ minimum input resistance

INTRODUCTION

The resistor and voltmeter in Fig. 13-6 are used to indicate the current in the circuit. The impedance of the parallel resonant *LC* is very high compared with the resistance of the resistor. Therefore, the current of the total circuit will be essentially determined by the resonant *LC* circuit. At resonance, the current in the circuit (and thus the voltage across *R*) will be at its minimum value.

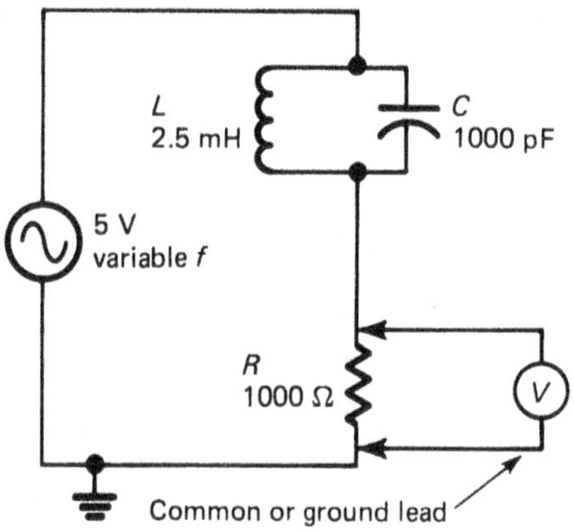

Fig. 13-6 Circuit for experiment 13-2, steps 1 and 2.

When the generator's frequency is at the edge of the bandwidth of the *LC* circuit, the current will be 1.414 times greater than at resonance. (This is because the impedance of the *LC* circuit is only 0.707 times as great as it was at resonance.) Therefore, the edge of the bandwidth of the *LC* circuit occurs when the voltage across *R* is 1.414 times as great as it was at resonance.

PROCEDURE

1. **a.** Calculate and record the resonant frequency of the circuit in Fig. 13-6.

   ____

   b. Construct the circuit in Fig. 13-6, and adjust the generator for 5-V output at the calculated resonant frequency. (Be sure to keep the ground terminal of the voltmeter connected to the ground terminal of the generator for all voltage measurements.) Now connect the voltmeter across the resistor. Carefully adjust the generator frequency to obtain the lowest possible voltage across the resistor. Record this voltage.

   ____

c. Determine and record the circuit current by the formula

$$I_T = I_R = \frac{V_R}{R} = \frac{V_R}{1000} = \_\_\_\_\_\_\_\_$$

d. Now determine and record the circuit impedance by the formula

$$Z = \frac{V_T}{I_T} = \frac{5}{I_T} = \_\_\_\_\_\_\_\_$$

e. Since Z is so high compared with R, you can assume that this calculated Z is the impedance of the LC part of the circuit. Does a parallel resonant circuit offer minimum or maximum opposition at its resonant frequency? ________

2. a. Slowly decrease the generator frequency until the voltmeter indicates 1.414 times as much voltage as it did in step 1b. This is the lowest frequency of the bandwidth; record it. ________

b. Now increase the frequency of the generator until V_R is again 1.414 times as great as it was in step 1b. Record the highest frequency of the bandwidth. ________

c. What is the bandwidth of the circuit? ________

d. Now determine and record the quality of the circuit by the formula

$$Q = \frac{f_r}{\text{BW}} = \_\_\_\_\_\_\_\_$$

3. Construct the series LC circuit of Fig. 13-7(a). Notice that the L and C are the same values you used in the parallel circuit.

a. What should the resonant frequency of the circuit be? ________

b. The resonant frequency of the circuit in Fig. 13-7(a) is that frequency which produces maximum indication on the voltmeter. Adjust the generator frequency for maximum voltage across the capacitor in Fig. 13-7(a). Record this voltage. ________

c. Now measure the voltage output from the generator. The generator voltage should be at its minimum value when the circuit is at resonance. Since the voltmeter has a small amount of internal capacitance, it may change the resonant frequency when measuring the total voltage. Therefore, readjust the generator's frequency to obtain the lowest possible indication on the voltmeter. Record this value.

____

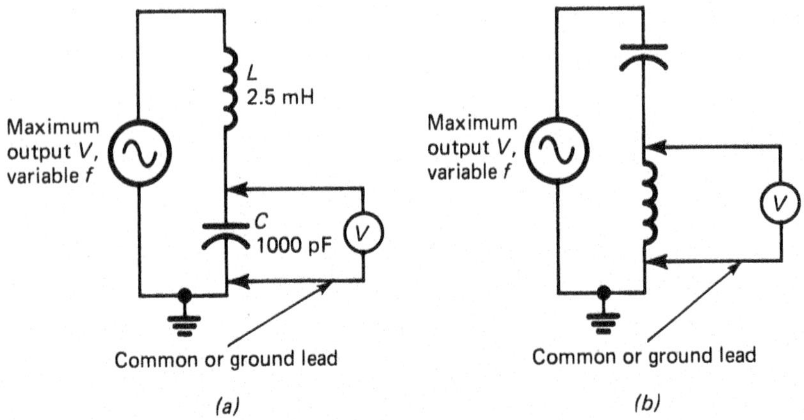

Fig. 13-7 Circuit for experiment 13-2, step 3.

d. Now you can calculate and record the quality of the circuit with the formula

$$Q = \frac{V_C}{V_T} = \_\_\_\_\_\_\_\_\_\_$$

e. Should this value of Q be the same as the value you calculated in step 2d? ______________

f. At resonance, should $V_C = V_L$? ______________

g. Check your response to step 3f by reversing L and C as shown in Fig. 13-7(*b*). Again adjust the frequency of the generator for maximum voltage across the inductor. Does $V_L = V_C$ at resonance? ______________

4. a. Calculate and record the reactance of the inductor in Fig. 13-7 at the resonant frequency of the circuit ($X_L = 6.28fL$). ______________

b. Now, using the value of Q from step 3d, calculate the effective resistance of the circuit ($R = X_L/Q$). ______________

c. How does this calculated value of R compare with the dc resistance of the inductor? ______________

d. Do your results indicate that the quality of a circuit is determined primarily by the resistance of the inductor? ______________

13-3 LAB EXPERIMENT: FILTERS 1

PURPOSE

The purpose of this experiment is to determine the characteristics of various filters. The experiment will also provide experience in plotting frequency response curves.

MATERIALS

Qty.

1 signal generator, audio to 150-kHz range
1 capacitor, 0.47 μF, $\pm$10%, 100 WV dc

Qty.

1 resistor, 220 Ω, $\pm$10%, ½ W
1 VOM or DMM

INTRODUCTION

Frequency response curves are usually plotted on graph paper that has a nonlinear scale on the horizontal axis. The nonlinear scale is called a logarithmic scale or a "log" scale. We will use this type of graph so that we can get a wide frequency range on a relatively small graph.

PROCEDURE

1. **a.** What type of filter circuit is shown in Fig. 13-8(*a*)? ____________
   b. Should the output voltage of the filter be greater at 100 or 600 Hz?
   ____
   c. Construct the circuit in Fig. 13-8(*a*), and adjust the generator frequency for 30 Hz. Adjust the generator's output voltage control to provide 5 V at the input terminals of the filter. Record the output voltage in the data table of Fig. 13-8(*b*). Does your answer to step 1a appear to be correct?
   ____

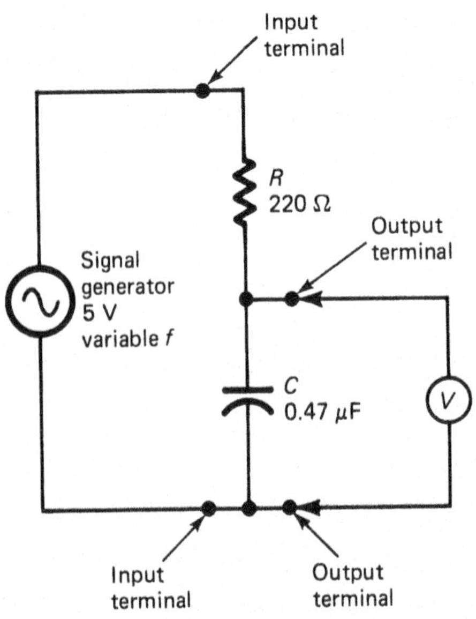

(*a*)

Data table

Frequency (hertz)	Output voltage (volts)
30	
60	
100	
300	
600	
1000	
3000	
6000	
10,000	
30,000	
60,000	
100,000	

(*b*)

Fig. 13-8 Circuit and data table for experiment 13-3, step 1.

Now change the generator frequency to 60 Hz, and adjust the generator output voltage (if necessary) to maintain 5 V input to the filter. Record the output voltage of the filter in the data table. Continue to change the frequency, adjust the generator output voltage, and record the filter output voltage until the data table is completed.

d. Use the data from Fig. 13-8(*b*) to plot frequency-voltage points on the semilogarithmic graph paper in Fig. 13-9(*a*). After you have plotted all 12 frequency-voltage points on the graph, connect them together to form a frequency response curve. From your completed curve, determine and record the filter output voltage at 800 Hz when the input is 5 V. ______________

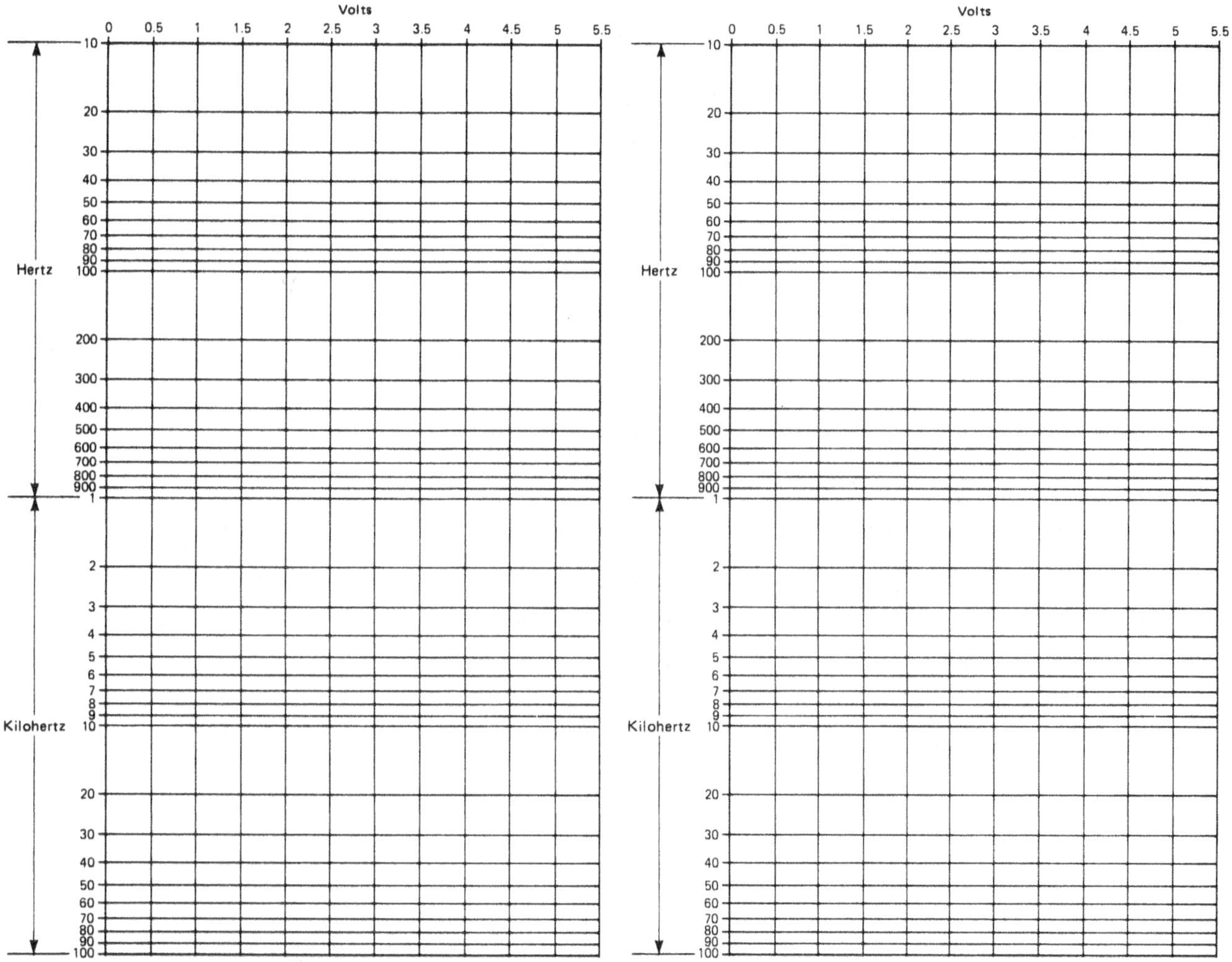

(*a*) (*b*)

Fig. 13-9 Semilogarithmic graph paper for experiment 13-3, steps 1d and 2d.

2. a. Refer to Fig. 13-10(*a*). Is this a high-pass or a low-pass filter? ______________

b. Should the filter output be smaller at 100 or 10,000 Hz? ______________

c. Construct the circuit of Fig. 13-10(*a*). Following the procedure used in step 1, complete the data table [Fig. 13-10(*b*)]. Do your data support your answers to steps 2a and b? ______________

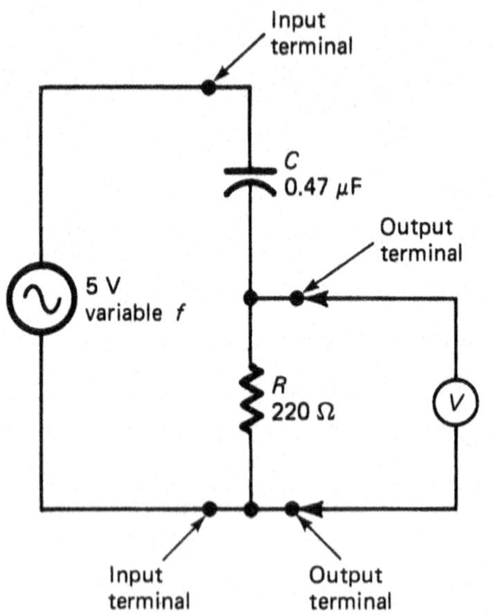

(a)

Data table

Frequency (hertz)	Output voltage (volts)
30	
60	
100	
300	
600	
1000	
3000	
6000	
10,000	
30,000	
60,000	
100,000	

(b)

Fig. 13-10 Circuit and data table for experiment 13-3, step 2.

d. Using the data collected in Fig. 13-10(*b*), construct a response curve on the semilogarithmic graph paper in Fig. 13-9(*b*). Compare the curves from steps 1d and 2d. What is the approximate frequency at which the two filters have the same output voltage? ______________

NAME ________________________________ DATE ________________

13-4 MULTISIM: FILTERS 2

PURPOSE

This Multisim experiment will provide you with experience in predicting the behavior of an *RC* filter. It also provides visual evidence of how a filter can repress an undesired signal voltage.

MATERIALS

Qty.

Multisim
Multisim file for Fig. 13-11

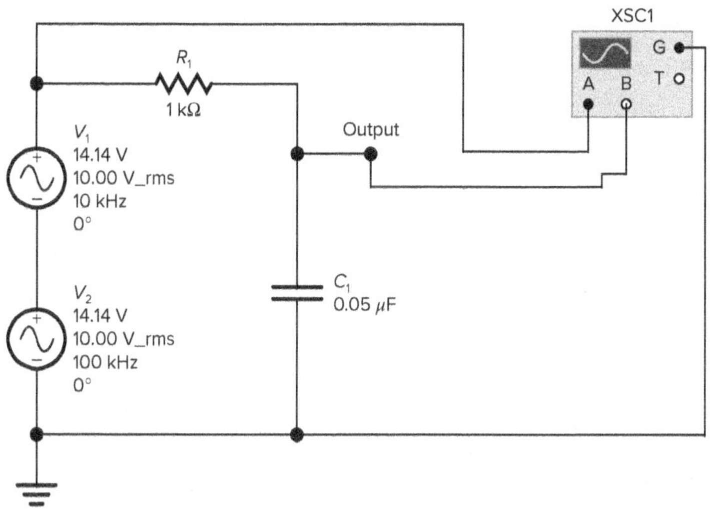

Fig. 13-11 Circuit for experiment 13-4.

PROCEDURE

1. What type of filter is shown in Fig. 13-11? ______________
2. Calculate and record the reactance of C_1 at each f used in Fig. 13-11.
   a. X_C at 10 kHz = ______________
   b. X_C at 100 kHz = ______________
3. What type of waveform will appear on channel A of the oscilloscope?

   ____
4. Will the low-frequency signal be significantly phase-shifted? ______________
   Why? ____
5. Activate the Multisim file for Fig. 13-11.
   a. Does Multisim support your answers to steps 1, 3, and 4 above?

   ____

   b. What should happen to the waveforms if R_1 is changed to 10 kΩ?

   ____

   c. Change R_1 to 10 kΩ and check your answer.
   d. What should happen to the waveforms if R_1 is changed to 500 Ω? ____

   ____

   e. Change R_1 to 500 kΩ and check your answer.

13-5 LAB EXPERIMENT: MEASURING PHASE RELATIONSHIPS

PURPOSE

The purpose of this experiment is to provide experience in measuring phase shift in an ac circuit. This experiment will provide further experience in using an oscilloscope.

MATERIALS

Qty.

- 1 oscilloscope with external trigger (or sync)
- 1 resistor, 2200 Ω, ±10%, ½ W
- 1 capacitor, 0.47 μF, ±10%, 100 WV dc
- 1 signal generator, sinusoidal output, audio range

INTRODUCTION

In experiment 8-3 you learned how to measure voltage (V) and frequency (f) with an oscilloscope. This experiment will use the oscilloscope in the same way as in experiment 8-3 except that you will use external triggering. To reacquaint yourself with the oscilloscope, reread the introduction to experiment 8-3. Then continue on with the next paragraph.

By using the external trigger of an oscilloscope, we can measure phase relationships of voltages in a circuit. External triggering causes the oscilloscope to respond to the time differences between the signals at the vertical input and external trigger input. The signal applied to the external trigger input becomes the reference signal. The phase angle of any signal applied to the main (vertical) input is referenced to the external trigger signal. This idea is illustrated in Fig. 13-12. In Fig. 13-12(*a*) three out-of-phase voltages are shown. They are all the same frequency. Suppose V_2 is applied to the external trigger input. Now, suppose we also apply V_2 to the main input. The result would be a waveform on the screen of the oscilloscope as in Fig. 13-12(*b*). The waveform would be the

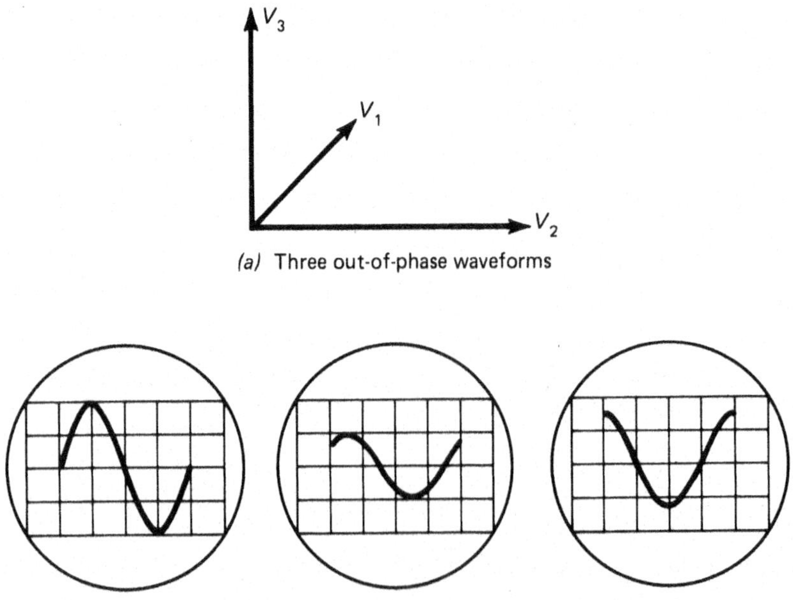

(*a*) Three out-of-phase waveforms

(*b*) V_2 displayed (*c*) V_1 displayed (*d*) V_3 displayed

Fig. 13-12 Phase relationships shown on oscilloscope.

reference waveform because it is a display of the voltage applied to the external trigger input. Notice in Fig. 13-12(*b*) that each division on the horizontal axis represents 90 electrical degrees.

Now suppose we remove V_2 from the main input and apply V_1 to the main input. (V_2, the reference signal, is still applied to the external trigger input.) V_1 will produce the waveform shown in Fig. 13-12(*c*). This waveform is clearly out of phase with the waveform in Fig. 13-12(*b*). Its peak value occurs 45° before the peak value of the waveform in Fig. 13-12(*b*). Also, notice that the waveform in Fig. 13-12(*c*) has less amplitude than the one in Fig. 13-12(*b*). In other words, the oscilloscope has, in two steps, displayed the voltage and phase relationships between V_1 and V_2 that are indicated in Fig. 13-12(*a*).

You can see from Fig. 13-12(*d*) what happens when V_3 replaces V_1 as the signal applied to the normal input. As expected from Fig 13-12(*a*), the waveform in Fig. 13-12(*d*) leads the waveform in Fig. 13-12(*b*) by exactly 90°.

If you are using a dual-trace oscilloscope, you can still follow the procedure discussed above. Just ignore the B channel and use only the A channel.

If, however, you elect to use both channels, select the A channel as the reference channel and leave it connected to the source voltage. Make the A channel the trigger source. Now you can use the B channel to measure V_R and V_C. You will still need to switch the position of R and C in order to measure V_R.

PROCEDURE

1. Construct the circuit shown in Fig. 13-13(*a*). Connect the main lead to the top of R (test point 1). Notice that V_T will be the reference voltage for this circuit as it is applied to the external trigger input.

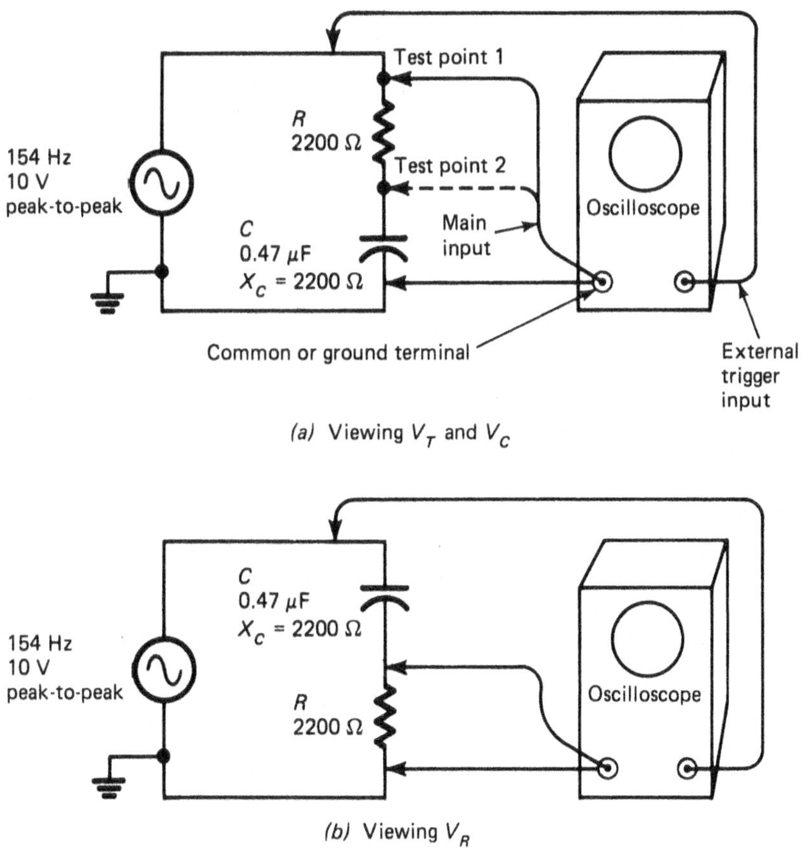

Fig. 13-13 Circuit for experiment 13-5.

Put the trigger selection control on the external trigger position. Adjust the trigger level control until you have a stable waveform. Then adjust the time/division controls until you have a single cycle displayed. Now adjust the output voltage of the generator for 10 V peak to peak (as measured by the oscilloscope). You now have one cycle of the 10-V peak-to-peak reference signal displayed. Center the waveform in the middle of the screen.

Change the main input connection from the top of *R* to the bottom of *R* (test point 2). This is shown by the dotted line connection in Fig. 13-13(*a*). (Do not adjust any of the controls on the oscilloscope.) The voltage V_C is now being displayed.

a. What is the peak-to-peak voltage across the capacitor? ______________

b. Does the capacitor voltage lead or lag the generator (reference) voltage? ______________

c. By what angle does the capacitor voltage lead or lag the reference voltage? ______________

d. Next, reverse the position of *R* and *C* in the circuit as shown in Fig. 13-13(*b*). You are now viewing the waveform of the resistor's voltage. What is the peak-to-peak voltage across the resistor? ______________

e. Does the resistor voltage lead or lag the capacitor voltage? ______________

f. By how many degrees does the resistor voltage lead or lag the capacitor voltage? ______________

g. Does the resistor voltage lead or lag the source voltage? ______________

h. By how many degrees does it lead or lag? ______________

2. a. Notice in Fig. 13-13 that *R* and X_C are equal. Therefore, their voltages should also be equal. Were the voltages measured in steps 1a and 1d within 10 percent of each other? ______________

b. According to Kirchhoff's voltage law, the instantaneous voltage of $V_C + V_R$ should equal V_T. Do the three voltages indicated by the waveforms on the screen of the oscilloscope verify this law? ______________

13-6 LAB EXPERIMENT: *LC* RESPONSE CURVES

PURPOSE

This experiment provides experience in using a response curve to determine f_r, BW, and Q. It also demonstrates how resistance determines the shape of the response curve and the Q of the circuit when C and L are unchanged.

MATERIALS

Qty.

Multisim
Multisim file for Fig. 13-14

INTRODUCTION

This experiment uses Multisim to display a series *LC* response curve (see Fig. 13-14) for a wide range of frequencies. By positioning the vertical cursor on the Bode plotter, you can determine the portion of the source voltage that develops across the current-sensing resistor at any frequency. The cursor can be moved by dragging it with the mouse or by using the direction arrows on the Bode plotter. The circuit, with the Bode plotter connected, is contained in the Multisim file for Fig. 13-14.

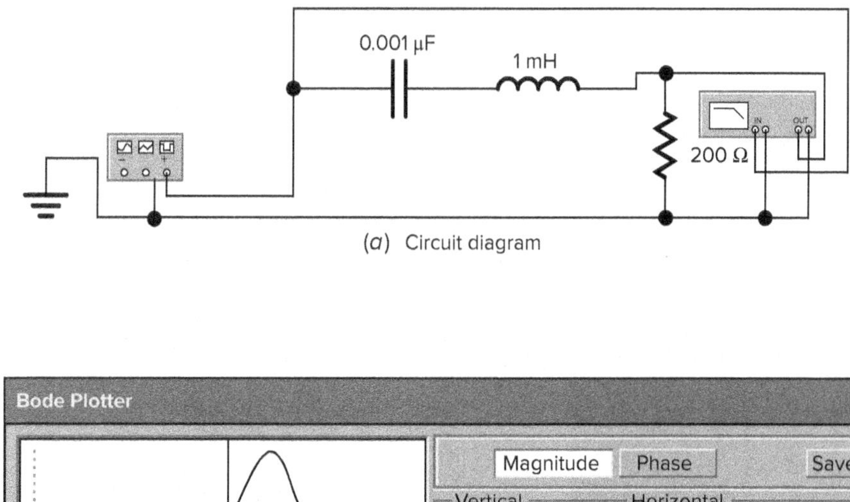

(*a*) Circuit diagram

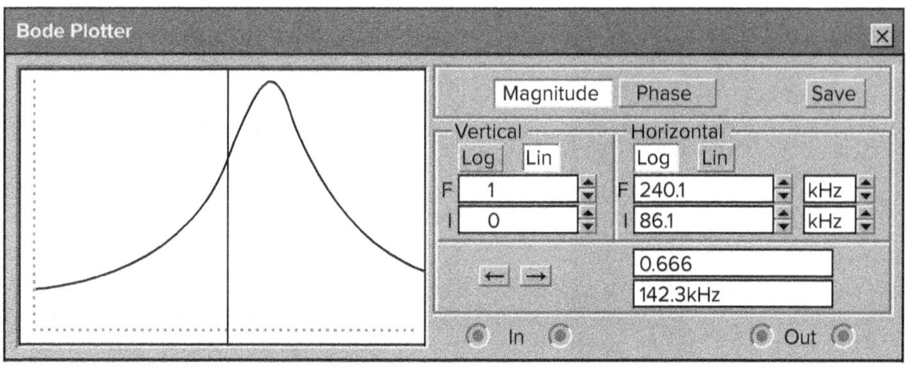

(*b*) Bode plotter settings and response curves

Fig. 13-14 Circuit for experiment 13-6.

PROCEDURE

1. Using Multisim, open the file for Fig. 13-14 and activate the circuit. Use the mouse to drag the vertical cursor to the approximate desired location. Then use the direction arrows for fine adjustment of the cursor location.

a. Determine and record the resonant frequency of the circuit by positioning the cursor in the center of the curve. ______________

b. Using the L and C values on the diagram, calculate and record the resonant frequency. ______________

c. Do the values obtained in steps 1a and 1b agree within 5 percent? ______________

d. Using the cursor, determine (and record) the upper and lower frequencies of the bandwidth.
Upper f ______________ Lower f ______________

e. Record the BW of the circuit. ______________

f. Calculate and record the Q of the circuit. ______________

g. If the resistance in the circuit is changed to 100 Ω, what value of Q and BW would you expect? Q ______________ BW ______________

h. Change R to 100 Ω. Now determine and record the new Q and BW.
Q ______________ BW ______________

i. Do Q and R appear to be directly or inversely proportional? ______________

j. Predict, and record your prediction for, the values of Q and BW when R is 20 Ω. Q ______________ BW ______________

k. Change R to 20 Ω and determine the new values for Q and BW.
Q ______________ BW ______________

l. What happened to the shape of the response curve as R was reduced?

____

____

NAME ______ DATE ______

13-7 LAB EXPERIMENT: TROUBLESHOOTING

PURPOSE

This experiment provides experience in analyzing and troubleshooting an *LCR* circuit. Assume the circuit components are soldered in place. Use only the *V* and *I* functions of the multimeter.

MATERIALS

Qty.

Multisim
Multisim file for Fig. 13-15

PROCEDURE

1. For the circuit in Fig. 13-15, calculate and record the following:
   X_C ______
   X_L ______
   Z ______
   I ______
   V_C ______
   V_L ______
   V_R ______
2. Open and activate the Multisim file for Fig. 13-15. Determine the faulty component. Submit a report detailing how you determined which component was faulty and the nature of the fault.

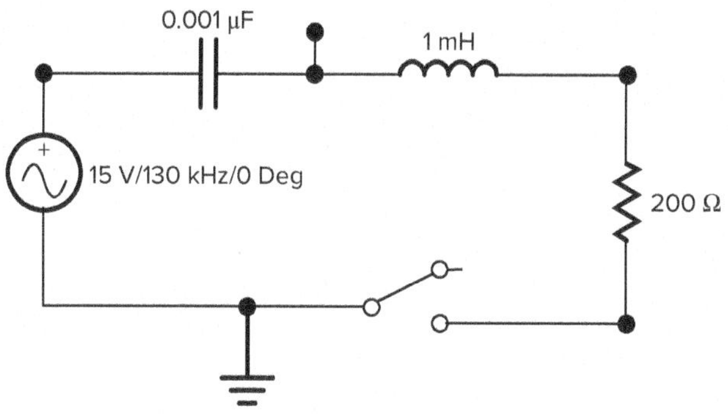

Fig. 13-15 Circuit for experiment 13-7.

NAME ______________________________ DATE ______________

CHAPTER 14

Electric Motors

TEST: ELECTRIC MOTORS

For questions 1 to 13, determine whether each statement is true or false.

1. A split-phase motor develops maximum torque at its rated speed. **1.** ______

2. A typical 240-V motor can safely operate on a nominal 230-V line that has long-term variations of ± 10 percent. **2.** ______

3. The wire in the start winding of a capacitor-start motor is larger than that in an equivalent hp split-phase motor. **3.** ______

4. Most permanent-capacitor motors use a salient-pole stator. **4.** ______

5. Split-phase motors are commonly available in ihp sizes. **5.** ______

6. "Pull-out torque" is another name for "locked-rotor torque." **6.** ______

7. The current and voltage that cause rotational torque in a squirrel-cage rotor are induced by generator action. **7.** ______

8. Slot insulators are used in the squirrel-cage rotor. **8.** ______

9. To optimize efficiency and PF, an induction motor should be operated at rated hp. **9.** ______

10. A waterproof motor prevents water from entering the motor enclosure under any operating conditions. **10.** ______

11. For optimum starting torque, the run-winding current should be 45° out of phase with the start-winding current. **11.** ______

12. All reluctance-start motors are synchronous motors. **12.** ______

13. The brushless dc motor is a variable speed motor. **13.** ______

For questions 14 to 20, choose the letter that best completes each statement.

14. The motor with the lower full-load current is the **14.** ______

a. Permanent-capacitor **c.** Split-phase
b. Split-capacitor **d.** Shaded-pole

15. Which of these motors uses a start switch? **15.** ______

a. Permanent-capacitor **c.** Shaded-pole
b. Capacitor-start **d.** Reluctance-start

16. The motor that has the greater starting torque is the **16.** ______

a. Reluctance-start **c.** Permanent-capacitor
b. Shaded-pole **d.** Capacitor-start

17. Which of these motors is synchronous? **17.** ______

a. Shaded-pole **c.** Permanent-capacitor
b. Reluctance **d.** Capacitor-start

18. Which of these motors is not easily reversible? **18.** ______

a. Two-value capacitor **c.** Shaded-pole
b. Split-phase **d.** Capacitor-start

19. Which motor receives dc pulses for its stator windings? 19. ____________

a. Split-phase c. Shaded-pole
b. Brushless d. Reluctance

20. Which motor requires a motor controller for its operation? 20. ____________

a. Two-value capacitor c. Split-phase
b. Reluctance d. Brushless dc

For questions 21 to 38, solve each problem. Be sure to include units (hp, rpm, etc.) in your answer when appropriate.

21. A motor has a rated speed of 1140 rpm and develops a torque of 2.88 lb-ft at rated speed. Determine its hp. 21. ____________
22. A 240-V, 60-Hz, 1725-rpm, ¾-hp motor has a service factor rating of 1.15. With a constant load, how much hp can it continuously deliver? 22. ____________
23. How much torque will a 120-V, 50-Hz, 1725-rpm, ½-hp motor develop at rated speed? 23. ____________
24. Determine the percentage of slip of a 120-V, 60-Hz, 4-pole, ¼-hp motor that operates at 1710 rpm. 24. ____________
25. What is the PF of a motor that is 70 percent efficient and draws 15 A from a 240-V source while delivering 2.5 hp? 25. ____________
26. What is the synchronous speed of an eight-pole, 100-V, 80-Hz motor? 26. ____________
27. A 440-V motor draws a full-load current of 12 A. It is to be connected to a source that provides a minimum of 400 V. What is the maximum distance the motor can be from the source if the resistance of the connecting conductors is 2.525 Ω/1000 ft? 27. ____________
28. Which type of dc motor has the better speed regulation? 28. ____________
29. Which type of dc motor has the greater starting torque? 29. ____________
30. Which type of dc motor should be used for a variable-speed operation? 30. ____________
31. Which type of compound winding (differential or cumulative) is usually used in a dc motor? 31. ____________
32. Which type of dc motor has electrical characteristics that are equivalent to those of the universal motor? 32. ____________
33. What are the names of the windings in dc and universal motors? 33. ____________
34. What type of dc motor provides good starting torque but doesn't reach a destructive speed when its load is disconnected? 34. ____________
35. At what percentage of rated speed does a start switch usually open? 35. ____________
36. A motor is rated for a 65°C temperature rise. What is its maximum operating temperature? 36. ____________
37. What is the efficiency of a motor that produces 0.75 hp when drawing 7.6 A at a PF of 0.8 from a 120-V source? 37. ____________
38. A 2-hp, 240-V motor has class B (130°C) insulation and can operate in an environment with 40°C ambient temperature. What is the temperature-rise rating of this motor? 38. ____________

NAME ____ DATE ____________________

14-1 LAB EXPERIMENT: REVERSING ROTOR ROTATION

PURPOSE

This experiment will provide experience in reversing the direction of rotation in a split-phase motor. In the process, you will see the construction details of a motor and learn to identify the run-winding and start-winding leads of a motor.

MATERIALS

Qty.

1 split-phase, four-pole, 115-V, ⅓- or ½-hp, reversible motor

1 multimeter

Qty.

• hand tools needed to disassemble and assemble the motor

CAUTION **Before connecting the motor to a 120-V outlet, be certain that (1) the motor is securely mounted on a stable base, (2) the motor is electrically grounded, and (3) there are no exposed electrical terminals or connections. Do not touch, or allow loose clothing or hair to touch, the rotating shaft of the motor.**

PROCEDURE

1. Apply power to the motor. Note, and record, whether the shaft rotates clockwise or counterclockwise. Remove power from the motor.
2. Before disassembling a motor, remove, with a file, any imperfections on the shaft that may mar the bearing surface when the motor is disassembled. Also, check the end shields and stator housing for alignment marks. If none are found, make alignment marks on these parts with a center punch. These marks are used in reassembling the motor so that parts will have their original mechanical alignment.
3. Following the procedure demonstrated by your instructor, disassemble the motor. Examine the centrifugal mechanism and locate the spring-retained weights. Pull out on these weights and note that the bobbin, which would normally be in contact with the start switch located in the end shield, moves back on the shaft toward the rotor. This is the action that opens the start switch.
4. Next, look at the start switch.
   - **a.** Is it open or closed when the motor is disassembled? ______________
   - **b.** Are the start windings connected in parallel with the run windings at this time? ______________ Why?

     ____

     ____
   - **c.** Measure and record the resistance between the power terminals on the terminal board. ______________
   - **d.** Is the resistance measured in step 4c the resistance of the run winding, the start winding, or the combination of the two windings? Note: The run winding, or both the run and the start winding, may be connected in series with a thermal cutout device, but the resistance of this device is very small compared with the resistance of the windings. Its resistance can be ignored in this activity.______________

e. Examine the stator of the motor and notice three things: the start winding is on the top of the run winding, the start-winding wire is much smaller than the run-winding wire, and the stator has four leads leaving it.

One of the four leads from the stator goes to the start switch, so it must be connected to one end of the start winding. Connect one lead of an ohmmeter (lowest range) to this lead. Two of the three remaining leads from the stator will be electrically connected together either by terminating on the same point (a terminal post) or by terminating on two points that are connected together by a jumper wire. Temporarily disconnect these two leads by removing one of them from a terminal post. Now, with the other lead from the ohmmeter, measure the resistance to each of the three isolated leads. The lead that yields the smallest resistance will, of course, be the other end of the start winding. (The resistance to the other two leads should be too great to measure with an ohmmeter.) Record the start-winding resistance. ______________

f. How does the start-winding resistance compare with the run-winding resistance?____

g. Finally, determine how you can reverse the leads to one of the windings without reversing the leads to both windings. On a motor that is made to be reversed, this can always be done by switching lead connections at the terminal board or at a junction box. Make the required lead changes and reassemble the motor.

Use the ohmmeter to check for shorts between the frame and the power leads. Of course, if there is a short, the cause must be determined and corrected.

Observing the caution (motor mounted, grounded, and insulated) at the beginning of this experiment, prepare to apply power but remember to *immediately* remove power if the motor does not operate properly. Does the motor now turn in the opposite direction from that recorded at the beginning of this experiment? ______________________

NAME ______________________ DATE ______________

14-2 LAB EXPERIMENT: MEASURING MOTOR CHARACTERISTICS

PURPOSE

This experiment will provide experience in using a clamp-on ammeter to measure and compare locked-rotor and no-load currents on two types of single-phase motors. Also, the locked-rotor torque of these two motors will be measured and compared.

MATERIALS

Qty.		Qty.	
1	clamp-on ammeter, multirange with a 100-A range	1	capacitor-start, four-pole, 115-V, ⅓- or ½-hp motor
1	split-phase, four-pole, 115-V, ⅓- or ½-hp motor	1	locked-rotor torque-measuring apparatus

CAUTION **Before applying power from the 120-V outlet, be certain that (1) the motor is securely mounted on a stable base, (2) the motor is electrically grounded, and (3) there are no exposed electrical terminals or connections. Do not touch the rotating shaft of a motor or let loose clothing or hair come in contact with the rotating shaft. Put the ammeter clamp around one of the current-carrying cord conductors before connecting the motor cord to the 120-V outlet.**

PROCEDURE

1. With the split-phase motor disconnected from the torque-measuring apparatus, measure and record in Fig. 14-1 its no-load current. Also, determine if the motor is rotating in the direction needed to measure its locked-rotor torque. If not, reverse the direction of rotation.

Type of Motor	Current			Torque	
	No-load	Locked-rotor	Rated	Locked-rotor	Rated
Split-phase					
Capacitor-start					

Fig. 14-1 Table for experiment 14-2, steps, 1, 2, and 3.

2. Connect the split-phase motor to the starting-torque-measuring apparatus. Also, connect the ammeter (100-A range). Have your instructor check your setup before applying power. To minimize the time that power is applied to the motor with a locked rotor, have one person read the ammeter while another person reads the force scale. *Do not* apply power for more than the second that it takes to read the instruments.
   a. Apply power; read the instruments; turn off power. Record the measured current and the torque in Fig. 14-1. (Torque is the force in pounds times the length in feet of the torque arm.)

b. Why is the locked-rotor current so much larger than the no-load current?

____

____

c. From the motor nameplate, obtain and record in Fig. 14-1 the rated current for the motor. Why is there such a small difference between the no-load and full-load (rated) currents?

____

____

d. From the nameplate data, determine and record in Fig. 14-1 the rated (full-load) torque of this motor (torque = 5252 hp/rpm). Is the locked-rotor torque greater than the rated torque? ____________ Should it be? ____________

3. Repeat steps 1 and 2 using the capacitor-start motor. Figure 14-1 should now be complete.

4. a. Which of the two motors you tested has the larger starting current? ____________ Why?

____

____

b. Which of the two motors has the larger starting torque? ____________ Why?

____

____

c. What difference, if any, does there appear to be in the run windings of these two motors?

____

____

d. Why were the motors operated for only a short period of time with locked rotors?

____

____

e. Would you expect the locked-rotor current to increase or decrease if power were applied for several seconds? ____________ Why?

____

____

NAME ______________________________ DATE ______________

14-3 MULTISIM: SHUNT MOTOR CHARACTERISTICS

PURPOSE

This simulation experiment demonstrates the major characteristics of a dc shunt motor. You will gain experience in determining I^2R loss, P_{out}, and efficiency using measured data gathered under a variety of load conditions.

MATERIALS

Qty.

Multisim
Multisim files for Fig.s 14-2 and 14-3

TABLE 14-1 Data from Simulation 14-3.

	Load Torque in Newton-Meters							
	N.L.	1.0	2.0	3.0	4.0	4.917	6.0	7.0
I_{arm}, A								
I_{field}, A								
S, rpm								
P_{in}, W								
I^2R_{field}, W								
I^2R_{arm}, W								
P_{out}, W								
eff, %								

INTRODUCTION

In the motor circuit provided in the Multisim file for Fig. 14-2, the voltmeter connected to the shaft of the motor is acting as a tachometer. Each volt registered on the meter represents one revolution per minute. Thus, a reading of 1.428 kV represents a shaft speed of 1428 rpm.

The torque load on the motor is expressed in newton-meters (Nm). One newton-meter of torque equals 0.7376 lb-ft.

The motor model used in Multisim assumes that core loss is negligible. The bearing friction is set so low that it is also negligible. Thus, the only losses to account for are the I^2R losses of the windings.

PROCEDURE

1. Using Multisim, open the file for Fig. 14-2. If a "Models clash" window appears, click on the "Use circuit model" button. Double-click on the motor symbol. Now click on the "Value" tab to display the characteristics of the motor used in this simulation experiment.
   a. Notice, and record, the following:
      Rated speed = ______________
      Rated armature I = ______________
      Rated source V = ______________

b. Also record the specified resistances for the armature and field coils.

R_{arm} ______________

R_{field} ______________

c. Notice the last entry (Load Torque) on "Value" listings. This is the only value you should change during the experiment. Enter zero for load torque and then click "OK." Finally, activate the circuit and record measured values in the first three rows of the N.L. column of Table 14-1. Why is the value of I_{arm} so very small?

____

____

d. Change the torque load to the next value called for in Table 14-1 by doing the following:

- Double-click on the motor symbol
- Click on the "value" tab
- Enter the desired load torque
- Click "OK"

Fill in the first three rows of the next column in Table 14-1.

e. Repeat step d until the first three rows of Table 14-1 are completed.

f. What happens to the speed when the load is increased?

____

____

g. Does the field current change as the load is changed? ______________ Why?

____

____

h. When the load torque is doubled, what happens to the armature current?

____

____

i. When the load torque is doubled, does the output power also double? ______________ Why?

____

____

j. Using your measured and recorded data, complete Table 14-1. P_{out} in watts can be calculated with the formula $P = 0.1048\ TS$, where T is in newton-meters and S is in revolutions per minute.

k. When does maximum efficiency occur for this motor?

____

l. What is the rated horsepower of this motor? (Remember: 1 hp = 746 W.)

____

m. Do the data in the P_{out} row of Table 14-1 support your answer to step i?

____

n. Do the power losses in the winding equal the difference between P_{in} and P_{out} for each load condition? ______________ Why?

____

____

2. Open and activate the Multisim file for Fig. 14-3. This is the same motor as in Fig. 14-2. It has a 1-Nm load torque.

a. Note and record the speed and currents.

I_{arm} = ______________

I_{field} = ______________

Speed = ______________

b. Should decreasing the resistance of the rheostat increase or decrease the speed? ____________ Why?

____

____

c. Decrease the resistance (use the r key) to 250 Ω and record the following:
I_{arm} = ____________
I_{field} = ____________
Speed = ____________

d. Decrease the resistance to 0 Ω. Do the meter readings now agree with those in column 2 of Table 14-1? ____________

e. With this 1-Nm load torque, what speed range was achieved without exceeding rated I_{arm}? ____________

f. Would you expect this same speed range with a 3-Nm load torque? ____________ Why?

____

____

g. Change the load torque to 3 Nm. Now determine and record the speed range obtainable without exceeding rated I_{arm}. ____________

h. Adjust the resistance to 15 percent (75 Ω). For this resistance setting, and the 3-Nm load torque, determine the following:
I^2R_{field} = ____________
I^2R_{arm} = ____________
P_{in} = ____________
P_{out} = ____________

i. Do the I^2R losses account for the difference between P_{in} and P_{out}?

____

j. How much speed range would you expect with a 4.917-Nm load torque? ____________ Why?

____

____

For questions 25 to 36, choose the letter that best completes each statement.

25. The meter movement that a rectifier-type meter uses is the **25.** ____________
a. Electrodynamometer
b. d'Arsonval
c. Vibrating-reed
d. Iron-vane

26. The thermocouple meter is used to measure **26.** ____________
a. Resistance
b. Current
c. Inductance
d. Capacitance

27. A multirange ammeter uses a switch that is **27.** ____________
a. Normally closed
b. Normally open
c. Shorting
d. Nonshorting

28. A multirange voltmeter uses a switch that is **28.** ____________
a. Normally closed
b. Normally open
c. Shorting
d. Nonshorting

29. To keep meter loading to less than 5 percent when measuring voltage across a 5-kΩ series resistor, the voltmeter input resistance should be at least **29.** ____________
a. 25 kΩ **c.** 500 kΩ
b. 100 kΩ **d.** 1 MΩ

30. The meter movement that uses a permanent magnet is the **30.** ____________
a. Electrodynamometer
b. d'Arsonval
c. Vibrating-reed
d. Iron-vane

31. The meter movement that uses more than one coil is the **31.** ____________
a. Electrodynamometer
b. d'Arsonval
c. Vibrating-reed
d. Iron-vane

32. The meter movement that is used in a wattmeter is the **32.** ____________
a. Electrodynamometer
b. d'Arsonval
c. Vibrating-reed
d. Iron-vane

33. The rheostat in an ohmmeter is used to **33.** ____________
a. Adjust for zero current in the meter
b. Adjust for the tolerance of the series resistor
c. Adjust for full-scale current in the meter
d. Adjust for infinity ohms

34. Increasing the voltage (and changing the series resistor) in an ohmmeter will **34.** ____________
a. Raise the center-scale resistance of the ohmmeter
b. Lower the center-scale resistance of the ohmmeter
c. Cause more current to flow in the ohmmeter
d. Cause less current to flow in the ohmmeter

35. Adding a shunt (and changing the series resistor) in an ohmmeter will **35.** ____________
a. Not change the center-scale resistance of the ohmmeter
b. Raise the center-scale resistance of the ohmmeter
c. Lower the center-scale resistance of the ohmmeter
d. Cause less current to flow in the ohmmeter

NAME ______________________________ DATE ______________

36. The meter used in a Wheatstone bridge is
a. An ammeter
b. A voltmeter
c. A d'Arsonval
d. A galvanometer

36. ______

For questions 37 to 47, solve each problem. Be sure to include units when appropriate.

37. A meter movement is rated at 50 μA and 250 mV. What is its internal resistance? **37.** ______

38. The power in a balanced three-phase load is being measured. A wattmeter indicates 1100 W for phase 2. What is the power of the load? **38.** ______

39. Two wattmeters are connected to a three-phase load. One wattmeter, with reverse polarity, indicates 360 W. The other wattmeter, with normal polarity, indicates 3200 W. What power is being used by the load? **39.** ______

40. Refer to Fig. 15-1. Find the value of C_u if R_2 is 4.2 kΩ when the bridge is balanced. **40.** ______

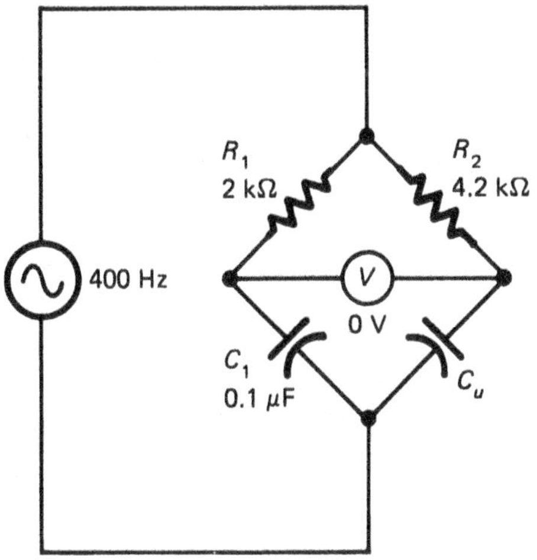

Fig. 15-1 Circuit for question 40.

41. What is the center-scale resistance of an ohmmeter that uses a 6-V battery and a 3-mA meter movement? **41.** ______

42. A bridge has a 2000-Ω resistor at R_1 and a 500-Ω resistor at R_2. It balances when the rheostat (replacing C_u in Fig. 15-1) is adjusted to 180 Ω. What is the value of the unknown resistor? **42.** ______

43. A 5000-Ω, 50-μA meter movement is used in a 1-mA ammeter. What is the resistance of the shunt? **43.** ______

44. What is the sensitivity (ohms-per-volt rating) of a voltmeter that uses a 60-μA meter movement? **44.** ______

45. What is the input resistance of a 20,000-Ω/V voltmeter on the 50-V range? **45.** ______

46. What value of multiplier resistance is needed to make a 25-V voltmeter from a 200-μA, 500-Ω meter movement? **46.** ______

47. What quantity is measured by a digital meter when measuring the value of each of the following components?
a. Resistor
b. Capacitor
c. Inductor

47. *Place answers below.*
a. ______
b. ______
c. ______

NAME ______________________________ DATE ______________

15-1 LAB EXPERIMENT: EXTENDING THE RANGE OF A VOLTMETER

PURPOSE

This experiment is designed to provide experience in working with voltmeters and multipliers. It will show how to extend the range of a panel meter with an external multiplier. It will also provide practical experience in using series-circuit relationships and Ohm's law.

MATERIALS

Qty.		Qty.	
1	dc panel meter, 5 V, 1000 Ω/V	1	resistor, 20,000 Ω, 1%, ½ W
1	dc power supply, variable, 0 to 20 V or 0 to 25 V	1	voltmeter (VOM or DMM), 25-V range

PROCEDURE

1. The range of the 5-V panel meter that has a sensitivity of 1000 Ω/V is to be extended to 25 V. To do this, you will need to add an external multiplier. The external multiplier's resistance is easy to find. The external multiplier's resistance will have to equal the difference between the input resistance of the 25-V meter and the input resistance of the 5-V meter.
2. Determine the total internal resistance (input resistance) of the 5-V panel meter. (This will include the meter-movement resistance plus the internal-multiplier resistance.)
   a. What is the internal resistance of the panel meter? ______________
   b. Next, calculate the input resistance that the 25-V meter will have. Note that the 25-V meter will also have a sensitivity of 1000 Ω/V. ______________
   c. Now, determine the resistance of the external multiplier needed to convert the 5-V meter to a 25-V meter. Connect this value of multiplier resistor in series with the 5-V panel meter, and you have a 25-V meter. ______________
3. Check the accuracy of your extended-range voltmeter by connecting the circuit shown in Fig. 15-2. Complete Table 15-1. The percentage of error is determined as follows:

$$\% \text{ error} = \frac{\text{VOM reading} - \text{extended range reading}}{\text{VOM reading}} \times 100$$

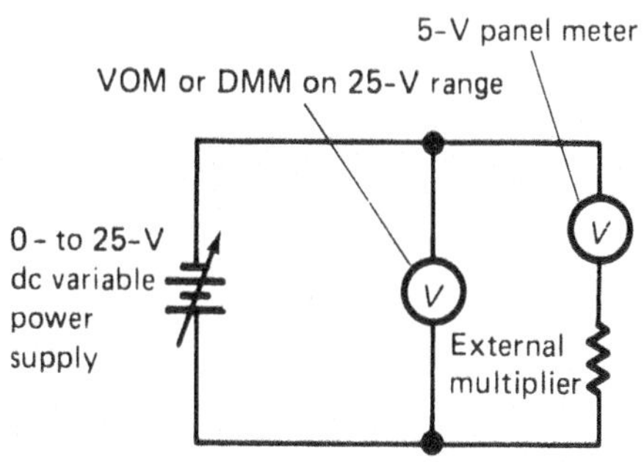

Fig. 15-2 Comparing voltmeter readings.

TABLE 15-1 Voltmeter Comparisons

VOM Reading	Extended-Range Voltmeter Reading	% Error
5 V		
10 V		
15 V		
20 V		
25 V		

NAME ______________________________ DATE ______________

15-2 LAB EXPERIMENT: VOLTMETER LOADING

PURPOSE

This experiment will illustrate voltmeter loading. It will show when it is necessary to use a DMM rather than a 20-kΩ/V VOM.

MATERIALS

Qty.

2 resistors, 1 kΩ, 5%, ½ W
2 resistors, 200 kΩ, 5%, ½ W
2 resistors, 10 MΩ, 5%, ½ W
1 dc power supply, 0 to 20 V

Qty.

1 DMM, at least 10-MΩ input resistance
1 VOM, 20-kΩ/V sensitivity

PROCEDURE

1. Refer to the circuit in Fig. 15-3(*a*).

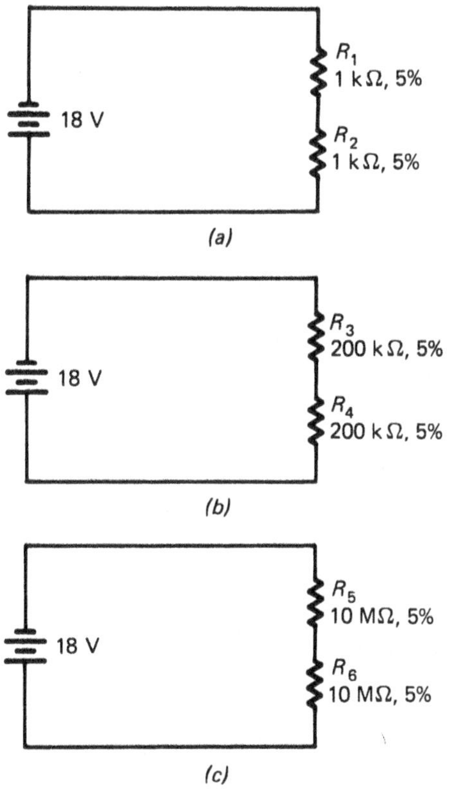

Fig. 15-3 Circuits for checking meter loading.

a. How much voltage would you expect across R_1? ______________
b. How much voltage would you expect across R_2? ______________
c. Would you expect voltmeter loading to occur in this circuit when voltages are measured with the DMM? ______________
The VOM? ______________ Why?

____

____

d. Construct the circuit in Fig. 15-3(*a*) and adjust the source voltage for exactly 18 V. Measure and record the voltages across R_1 and R_2 with the VOM on the 10-V range.

V_{R_1} = ______________ V_{R_2} = ______________

e. Next, measure and record these voltages with the DMM on the 10-V range.

V_{R_1} = ______________ V_{R_2} = ______________

f. Allowing for tolerances of the resistors and accuracy of the meters, do the measured voltages agree with the predicted voltages?

____

g. Did voltmeter loading occur in this circuit? ______________

2. Notice that the series resistors in Fig. 15-3(*b*) are equal, as they were in Fig. 15-3(*a*). Therefore, each resistor will still drop 9 V.

a. Do you predict voltmeter loading will occur in this circuit when using the DMM? ______________ The VOM? ______________ Why?

____

____

b. Construct the circuit in Fig. 15-3(*b*). Measure and record the voltages across the two resistors with the VOM and the DMM. With the VOM on the 10-V range, what were the voltages for V_{R_3} and V_{R_4}?

V_{R_3} = ______________ V_{R_4} = ______________

c. With the DMM, what were the readings for V_{R_3} and V_{R_4}?

V_{R_3} = ______________ V_{R_4} = ______________

d. Which meter, the VOM or the DMM, gave the more accurate measurements? ______________

e. Using the voltages measured with the VOM, does $V_T = V_{R_3} + V_{R_4}$?

____

f. Will the measured voltages in a series circuit satisfy Kirchhoff's voltage law when meter loading occurs? ______________

3. Construct the circuit of Fig. 15-3(*c*). If necessary, readjust the dc power supply for exactly 18 V. Measure and record V_{R_5} and V_{R_6} with each of the two meters. (Use the 10-V range on each meter.)

a. With the VOM, what were the measured voltages for V_{R_5} and V_{R_6}?

V_{R_5} = ______________ V_{R_6} = ______________

b. With the DMM, what were V_{R_5} and V_{R_6}?

V_{R_5} = ______________ V_{R_6} = ______________

c. Did meter loading occur with both meters? ______________

d. Which meter yielded the more accurate results? ______________

e. Knowing the input resistance of the DMM, could you have predicted the voltage indicated by the DMM across R_5? ______________ How?

____

____

NAME ______________________________ DATE ______________

15-3 LAB EXPERIMENT: DC BRIDGE

PURPOSE

This experiment provides experience in using a Wheatstone bridge to measure a wide range of resistances. You will learn how to change the range of a bridge.

MATERIALS

Qty.

1 DMM
1 potentiometer, 0 to 10 kΩ, 2 W, 5%, linear taper
1 resistor, 220 Ω, 10%, ½ W
1 resistor, 470 Ω, 10%, ½ W
1 resistor, 1000 Ω, 2%, ½ W
1 resistor, 10 kΩ, 2%, ½ W
1 resistor, 20 kΩ, 1%, ½ W

Qty.

• Assorted test resistors between 10 Ω and 20 kΩ
1 6-V dc supply
1 switch SPDT toggle
or
Multisim
Multisim file for Fig. 15-4

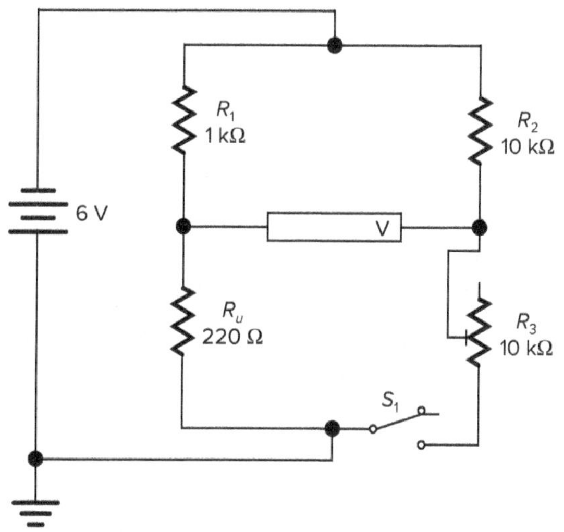

Fig. 15-4 Circuit for experiment 15-3.

PROCEDURE

Note: If you are using Multisim, S_1 is not necessary. You do not need to measure the value of R_1; its value can be figured from its percent setting when the bridge is nulled. The DMM is autoranging, so you need not take precautions to avoid overloading the meter. The circuit you will use is located in the Multisim file for Fig. 15-4.

1. Construct the circuit shown in Fig. 15-4. Set the DMM on a range greater than 6 V. Close S_1 and apply power to the circuit.

a. At what value of R_3 would you expect the bridge to be nulled? ______________

b. Adjust R_3 for null (lowest reading on the meter). If you reduce the range of the meter to obtain a more precise null, be sure to return it to a range greater than 6 V before opening S_1. Open S_1 and measure (and record) the value of R_3. ______________

c. Assuming that R_3 can be accurately adjusted from 100 to 9800 Ω, calculate and record the minimum and maximum resistance that the bridge can measure with $R_1 = 1$ kΩ and $R_2 = 10$ kΩ.

$R_{min} =$ ______________ $R_{max} =$ ______________

d. Check your answers to step c by measuring available resistors that are as close as possible to the values you calculated in step c. Do your answers to step c seem reasonable? ______________

e. Change R_1 to a 20-kΩ resistor. Assuming the same range for R_3 as in step c, calculate and record minimum and maximum values of R_u that can be measured.

$R_{min} =$ ______________ $R_{max} =$ ______________

f. Check your answers to step e by measuring available resistors that are as close as possible to the values you calculated in step e. Do your answers to step e seem reasonable? ______________

g. Change R_2 to 1 kΩ and leave R_1 at 20 kΩ. If R_3 can be adjusted to 10 kΩ, how large can R_u be and still achieve bridge balance (null)?

____

h. Change R_2 to 1 kΩ and check your answer to step g. Were you correct?

____

NAME ______________________________ DATE ______________

15-4 LAB EXPERIMENT: AC BRIDGE

PURPOSE

The purpose of this experiment is to construct an ac bridge. With the bridge you will determine the capacitance of several capacitors.

MATERIALS

Qty.

1 VOM or DMM
1 potentiometer, 0 to 10 kΩ, 2 W, 5%, linear taper
1 resistor, 1 kΩ, ±2%, 1 W
1 resistor, 10 kΩ, ±2%, 1 W
1 capacitor, 1000 pF, ±10%, 100 WV dc
1 capacitor, 1500 pF, ±10%, 100 WV dc
1 capacitor, 0.005 μF, ±10%, 100 WV dc

Qty.

1 capacitor, 0.01 μF, ±10%, 100 WV dc
1 signal generator, audio range
1 switch, SPST, 1 A, 125 V
or
Multisim
Multisim file for Fig. 15-5

PROCEDURE

Note: If you have a calibrated potentiometer, use it for R_1. Then when the directions call for measuring the value of R_1, you can just read the value of R_1 from the dial plate.

If you are using Multisim, S_1 is not necessary and you do not need to measure the value of R_1. Just figure its value from the percent setting when the bridge is nulled. The circuit you will use is located in the Multisim file for Fig. 15-5.

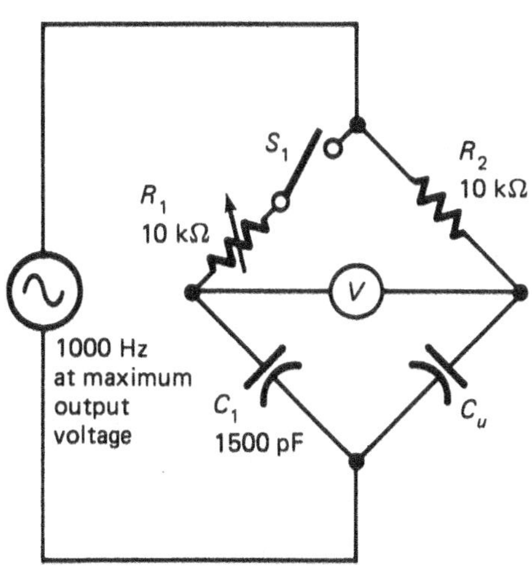

$$C_u = \left(\frac{C_1}{R_2}\right) R_1$$

Fig. 15-5 Circuit for experiment 15-4.

1. Construct the bridge circuit in Fig. 15-5, using a 1000-pF capacitor for C_u. Rearranging the bridge formula (Fig. 15-5) to solve for R_1 yields

$$R_1 = \frac{R_2 C_u}{C_1}$$

PROCEDURE

1. Refer to Fig. 15-10.
   a. What range is the ohmmeter set on? ______________
   b. What is the maximum resistance it can measure on this range? ______________
   c. What range would the voltmeter circuitry in the DMM be set on? ______________
   d. What readout would you expect when the simulation in Fig. 15-10 is activated and S_1 is closed? ______________
   e. Activate the Multisim file for Fig. 15-10.
   f. Was your answer to step 1d above correct (within acceptable tolerances)? ______________
   g. What readout would you expect when S_1 is open? ______________
   h. Open S_1 and check your answer to step 1g.
2. Refer to Fig. 15-11.
   a. How much current should be flowing in Fig. 15-11 when the Multisim file is activated? ______________
   b. What range is the ammeter set on?______________
   c. What readout would you expect when the circuit is activated? ______________
   d. Activate the Multisim file for Fig. 15-11.
   e. Was your answer to step 2c correct? ______________
   f. Does the ammeter cause any significant meter loading in this circuit? ______________

NAME ______________________________ DATE ______________

CHAPTER 16

Residential Wiring Concepts

TEST: RESIDENTIAL WIRING CONCEPTS

For questions 1 to 20, determine whether each statement is true or false.

1. A conductor with white insulation is never used as a hot conductor.
2. Three-phase power is often distributed throughout residential areas.
3. A conductor with black insulation is never used as a neutral conductor.
4. In some circuits, a white-colored conductor is connected to a black-colored conductor.
5. A neutral-gray-colored conductor can be used as a neutral conductor.
6. A white-colored conductor is never connected to a switch.
7. The energy meter in the service entrance provides continuity for the neutral conductor.
8. A distribution panel provides separate hot buses for 120-V and 240-V branch circuits.
9. Single-phase and three-phase power require separate lines for cross-country transfer of power.
10. A subpanel provides separate buses for neutral and equipment-grounding conductors.
11. A GFCI trips when the neutral-conductor current equals the hot-conductor current.
12. Most service entrances have a master disconnect switch as well as a main circuit breaker.
13. Most snap switches in residential wiring do not have grounding screws.
14. Removing the hot jumper bar from a duplex receptacle allows one of the receptacles to be switched.
15. A GFCI protects a circuit from excessively high load currents.
16. A 10-AWG conductor can safely carry more current in a 120-V circuit than in a 240-V circuit.
17. The term *service drop* refers to the three conductors from the power company's transformer that are connected to the service entrance.
18. Metal conduit is sometimes used as the equipment-grounding conductor.
19. The cable connecting two outlet boxes containing a three-way and a four-way switch will have three insulated conductors.
20. A toggling relay is used in all low-voltage-controlled lighting circuits.

For questions 21 to 34, supply the missing word or phrase needed in each statement.

21. The grounded conductor in a residential wiring system is called the ______________ conductor.
22. Controlling a light from four locations requires ______________ four-way switch(es).

23. A three-way switch has ______________ pole(s). 23. ______________
24. A four-way switch has ______________ throw(s). 24. ______________
25. A(n) ______________ cable connects to a circuit breaker at both the main panel and the subpanel. 25. ______________
26. The lighted-handle snap switch is illuminated when the controlled lamp is turned ______________. 26. ______________
27. Screws that receive equipment-grounding conductors are colored ______________. 27. ______________
28. A(n) ______________ disconnects a circuit before the ground-fault current can do damage to the equipment or operator. 28. ______________
29. Light-colored screws on a receptacle indicate the ______________ side of the receptacle. 29. ______________
30. The pole of a three-way switch is identified by a(n) ______________ colored screw. 30. ______________
31. A 10-AWG conductor can be connected to a circuit breaker rated at no more than ______________ amperes. 31. ______________
32. In the ______________ the neutral conductor is an uninsulated conductor. 32. ______________
33. The neutral conductor is also called the ______________ conductor. 33. ______________
34. In a low-voltage-controlled circuit, a remote switch is wired in ______________ with the local switch. 34. ______________

Answer questions 35 to 38 in the space provided.

35. What type of lamp is used in the handle of a lighted-handle snap switch? 35. ______________
36. Is a wire connector used on the neutral conductors in an outlet box that has two 12-2/G cables and a duplex receptacle? 36. ______________
37. Do both the neutral conductor and the hot conductor connect to the GFCI breaker? 37. ______________
38. What conductor (wire) connects to the neutral bus in a branch circuit protected by a GFCI breaker? 38. ______________

NAME ______________________________ DATE ______________

16-1 MULTISIM: TROUBLESHOOTING 1

PURPOSE

Troubleshooting a three-way switching circuit without disconnecting any conductors at the switches requires careful analysis of the circuit behavior and the voltage readings taken across the switch terminals. This faulty circuit requires such an analysis. The circuit diagram has extra dots added to it so that a voltmeter lead can be connected to any switch terminal or ground.

MATERIALS

Qty.

Multisim
Multisim file for Fig. 16-1

PROCEDURE

Using Multisim, open the file for Fig. 16-1 and activate the circuit. From observation of the circuit behavior and voltmeter readings, determine what the fault is—for example, short between throws or open between pole and throw. Submit a report detailing how you arrived at your conclusions.

16-2 MULTISIM: TROUBLESHOOTING 2

PURPOSE

Careful analysis of circuit behavior and voltage measurements (both expected and observed) is required in troubleshooting a three-way switching circuit. This faulty circuit requires such an analysis. The circuit diagram has extra dots added to it so that a voltmeter lead can be connected to any switch terminal or ground.

MATERIALS

Qty.

Multisim
Multisim file for Fig. 16-2

PROCEDURE

With Multisim, open the file for Fig. 16-2 and activate the circuit. From observation of the circuit behavior and voltmeter readings, determine what the fault is—for example, short between throws or open between pole and throw. Submit a report detailing how you arrived at your conclusions.

16-3 MULTISIM: TROUBLESHOOTING 3

PURPOSE

Troubleshooting this faulty three-way switching circuit requires careful analysis of circuit behavior and voltage measurements (both expected and observed). The circuit diagram has extra dots added to it so that a voltmeter lead can be connected to any switch terminal or ground.

MATERIALS

Qty.

Multisim
Multisim file for Fig. 16-3

PROCEDURE

With Multisim, open the file for Fig. 16-3 and activate the circuit. From observation of the circuit behavior and voltmeter readings, determine the fault—for example, a short between throws or an open between pole and throw. Submit a report detailing how you arrived at your conclusions.

16-4 MULTISIM: FOUR-WAY CIRCUIT

PURPOSE

This Multisim experiment will provide you with experience in wiring a four-way lighting circuit.

MATERIALS

Qty.

Multisim
Multisim file for Fig. 16-4

PROCEDURE

1. a. Open the Multisim file for Fig. 16-4. All of the components needed for the four-way lighting circuit are present. After you have connected the components together, activate the circuit and check to see that it works properly.

   b. Did your circuit operate properly? ____________________

   c. How many retries did it take to get the circuit functioning properly? ____________________

   d. Have your instructor check your circuit to verify that it is functioning properly.

NAME DATE

16-5 LAB EXERCISE: WIRING A FOUR-WAY CIRCUIT

PURPOSE

This exercise provides experience in determining the number of conductors required in the cables in a four-way circuit. It also requires determining which conductor should be connected to which terminal.

PROCEDURE

Complete the diagram in Fig. 16-1. Clearly label the color of the insulation on each conductor in each cable. Also show where wire connectors are needed and where connections are made to an outlet box. Use the codes given in Fig. 16-1.

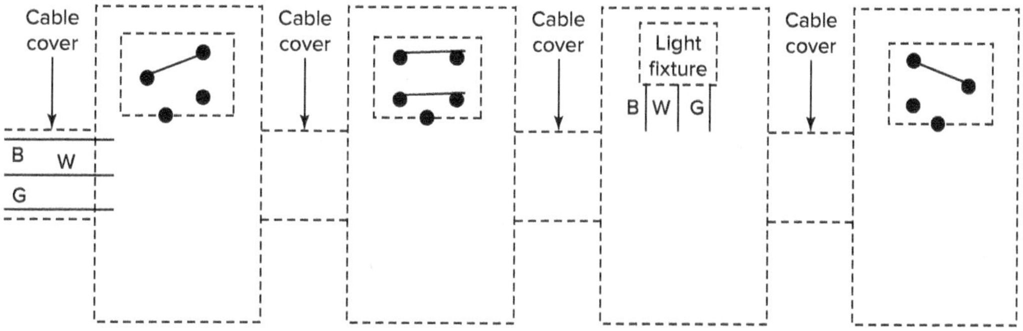

Code for insulation color
W = white
B = black
R = red
G = green or bare

Use a ● for a wire connector or a connection to a metall outlet box.

Fig. 16-1 Wiring a four-way circuit.

NAME ______________________________ DATE ______________

APPENDIX A

Breadboarding

CAUTION **Breadboarding is generally limited to low-power circuits. High currents would cause high temperatures, and the breadboard would be damaged. High voltage could cause arcing and would be very dangerous. *If there is any doubt, consult your instructor before breadboarding a circuit.***

The name *breadboarding* has its origin in the early years of electronics, when circuits were sometimes fashioned on a flat piece of wood that looked like a breadboard (or actually was a breadboard). In case you didn't know, breadboards are used in some kitchens for working with dough or cutting bread. Today, wood breadboards are seldom if ever used, and circuits are prototyped and tested using electronic breadboards, which are molded plastic with holes and internal metal contact strips. A circuit can be easily and quickly built by inserting component leads and jumper wires into the holes. Electronic breadboards are also called *prototyping boards*.

Figure A-1(*a*) shows how a series circuit appears on a typical electronic breadboard. Each group of five holes is electrically connected. Or, put another way, there is electrical continuity from any one contact to the other four contacts in that group. So, the top horizontal resistor is connected to the vertical resistor because each has a lead that is inserted into one group of five in Fig. A-1(*a*). Note that the vertical resistor and the bottom horizontal resistor are connected using another group of five.

There are also buses on breadboards. They are most often used for power, ground, or any signal requiring more than five common contact points. Figure A-1(*a*) shows two buses at the top and two at the bottom. Every contact along a bus is electrically connected to every other contact on that bus. Thus, the top source wire in Fig. A-1(*a*) is electrically connected to the top horizontal resistor via the top bus and jumper wire. Likewise, the bottom source wire is connected to the bottom horizontal resistor via the bottom bus and jumper wire. Note that the four buses in Fig. A-1(*a*) are electrically isolated from each other. So, one could use them as follows:

- +12-V bus
- −12-V bus
- +5-V bus
- Ground bus